The Evidential Details Mystery Series

Decorated United States Military Intelligence Psychic Remote Viewer solves some of History's Greatest Mysteries

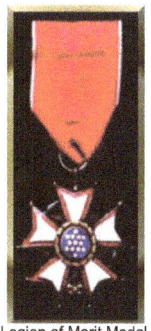
Legion of Merit Medal

Civil War Lost Order Mystery Solved

Seeds/McMoneagle

2016

The Logistics News Network, LLC. Chicago, Illinois

Evidential Details

e-mail

"Excellent books and research! Thanks for what you are doing."

"I'm not really all that clued on the civil war, however it was a great example of operational RV [Remote Viewing] by Joe [McMoneagle] ...and very well researched by yourself. I enjoyed it immensely. Looking forward to the next releases."

"I found the book to be very intriguing and enjoyed it. Thank you. The existence of such people [like McMoneagle] is quite extraordinary."

"Received your book yesterday and thought I'd look at it briefly as I had things to do. But once I opened it, I kept reading it. Amazing ...fascinating"

"Like the other two, Amelia [Earhart] and Princess Di, this is a fascinating story. The chapter was another example of the fact that you have a terrific idea, the historical mystery plus the answer to the puzzle per remote viewing."

"The book was excellent. Not too long and not too short either and your research was particularly very good."

"There is something about these stories you and Joe McMoneagle have done that has the ring of deep truth, and grabs the reader on a gut level. This is hard for me to explain, but I just wish you could find some way to get broader circulation of these wonderful projects."

"As time goes on others will get word and pick up the book(s)."

Evidential Details

The Quick Take

Once it was documented what this soldier could do, these Services submitted targets:

During his career, Mr. McMoneagle has provided ...informational support to the Central Intelligence Agency (CIA), Defense Intelligence Agency (DIA), National Security Agency (NSA), Drug Enforcement Agency (DEA), Secret Service, Federal Bureau of Investigation (FBI), United States Customs (ICE), the National Security Council (NSC), most major commands (Army, Navy, Air Force, Intelligence) within the Department of Defense (DOD), and hundreds of other individuals, companies, and corporations.

Paragraph from Mr. McMoneagle's CV.

Based on the quality of the Information, the Military's Decorations Committee selected the medal with this language:

The award is given for service rendered in a clearly exceptional manner. For service not related to actual war the term "key individual" applies to a narrower range of positions than in time of war and requires evidence of significant achievement. In peacetime, service should be in the nature of a special requirement or of an extremely difficult duty performed in an unprecedented and clearly exceptional manner.

The U.S. Army'ys Legion of Merit Medal bestowal prerequisites.

Evidential Details

Medals Received

Legion of Merit

Meritorious Service

Citations

Meritorious Service with **one** Oak Leaf Cluster;[1]
Army Commendation with **two** Oak Leaf Clusters, Presidential Unit;
Meritorious Unit with **three** Oak Leaf Clusters;
Vietnam Gallantry Cross with Palm for gathering enemy intelligence for Allied counter offensives.

---

[1] The same medal cannot be given twice. When exemplary service is again rendered, the Oak Leaf Cluster (OLC) is attached in place of a second identical medal.

Evidential Details

The Steel Plaque on the Brick Wall

In what is clearly the most fascinating component of U.S. Military History, Joseph McMoneagle is the only man in history to be awarded medals for consistent accuracy in Remote Viewing (Psy-functioning) by a military. As *Operation Star Gate's* Number 1 Military Intelligence asset at Fort Meade, Maryland, he was the Pentagon's go to man when secret data could not be obtained by any other means or was time sensitive.

Evidential Details

Published on the
Evidential Details Imprint,
a Division of Logistics News Network, LLC
P.O. Box 3600, Lisle, Illinois 60532

Revised Edition Copyright 2016 by LNN, llc.
RV Sessions ©1997 All rights reserved

The Evidential Detail's Mysteries Series

Civil War Lost Order Mystery Solved includes biographical references.
ISBN: 978-0-9826928-6-8

United States – History – Civil War, 1861-865 - Generals, United States – Civil War Campaigns – Civil War Correspondence – Battles, South Mountain, Harper's Ferry - 19th Century Military Intelligence - U.S. Military Intelligence - Remote Viewing – Joseph McMoneagle – Anomalous Cognition – Stanford Research Institute – Dr. Harold Puthoff

Library of Congress Card Number: 2011905178

Book layout and design by Logistics News Network, llc.
Printed in the United States of America

If you are unable to purchase our books in your bookstore,
they are available by e-mail at Manager@LNN1.com or
order from our web site at www.EvidentialDetails.com.

All rights reserved. No part of this book may be reproduced or transmitted in any form or by means, electronic or mechanical, including photo-copying or by any information storage and retrieval system without the written permission of the Publisher under International and Pan American Copyright Conventions

Note: Any interpretations or historical conclusions herein cannot be considered to represent the opinion of, or to be endorsed by, Joseph McMoneagle or any other individual. The final manuscript was not submitted for approval. There was no prior knowledge of when it was to be made available or the medium this research would be brought forward.

Pictures on the Internet and the Millennium Copywrite Act of 2000, on the heels of the Sony Bono Copyright Act, continue to be problematic. It remains an effort where organizations can only strive to be in compliance. In the event of inaccurate credit, please contact us with corrected information and proof of property ownership.

Word *only* portions, of this book may be reproduced for academic dissertations or in non-fiction works within lawfully established Fair Use guidelines with proper crediting. No text may be altered.

Evidential Details

Table of Contents

The Quick Take..5

Acknowledgment...10

Preface..11

Introduction: Former Princess Diana Spencer's 1997 auto accident......13

Part II

Target: Top Secret Confederate Order Lost……….....…....…………..45

Bibliography...110

Part III

Incredible Credentials..114

The Military's R.V. "Human Use" Clearances Conflict...............116

Encounters with the Chinese RV Program.................................117

The Army's Remote Viewing Project Protocols.........................119

Program Beginnings - Director Harold E. Puthoff, PhD............124

The Targeted Military Remote Viewers Bookshelf....................142

The Evidential Details Series Further Taskings........................144

Acknowledgement

The author would like to acknowledge the hours spent by Joseph McMoneagle. He is a peerless individual in the world and will be understood for generations as an exceptional scientific pioneer into the human brain and the probable co-incarnational condition of our three dimensional existence. I want to thank all the others in the original unit that I have met, or corresponded with, or telephoned, or studied under, or simply shook hands with.

I would like to acknowledge the people of the Monroe Institute™ whose pioneering programs do so much to help people understand the possibilities of the human mind.

The people of the National Park Service [Department of Interior] should also get an honorable mention for helping to identify the correct river bend mapped out by McMoneagle. I also want to thank Craig for his patience, support and feedback. And then, to all the positive and truly fascinated people, of every persuasion, whose judgment it was that this new data be brought forward.

Evidential Details

Preface

Many agree these books can affect history - for two reasons: findings and methodologies. For the first time, Military Intelligence quality Controlled Remote Viewing capabilities have been brought to solve historical mysteries in what academia refers to as *recent modern* – the exception being the Ötzi The Iceman material, which was designed to support scientific inquiry at the Museum in Bolzano, Italy. This effort was to develop the most accurate historical information possible. Time will tell as individuals, with peripheral research, provide anecdotal information or produce a document that suddenly fits.

    Others have gone so far as to say these findings unlock more new data than any particular PhD candidate's History dissertation this year. And it is for this prospect that the reader is encouraged to look deeper into the ulterior motives for opposition. For an indication as to the difficulties bringing this manuscript forward, the reader is referred to Jim Marr's Foreword in Lyn Buchanan's book *The Seventh Sense* [Paraview, ©2003] For information about the origins of the military's involvement with Remote Viewing, kindly refer to Beginnings. The Princess Diana Spencer Introduction was designed to answer questions about nomenclature, viewer integrity and data quality.

    For any researcher, it is the guarded realization that you have connected unattainable investigational dots that is the most intellectually stimulating factor in any horizon field of inquiry. My decade of documenting those points seem to confirm the viability of Remote Viewing as a research tool when used in conjunction with other research sources. What I can say is that I have been involved with the most spectacularly fascinating process of historical inquiry ever. As likely the only author to begin historical research with solutions in hand, I came to call these new confirmational dots the *Evidential Details*.

Evidential Details

"The worst term of all is "psychic." No stable definition has ever been established for it, and there are great hazards in attempting to utilize a term which has not much in the way of an agreed-upon definition.
　　Supporters do assume that it refers to extraordinary, non-normal (paranormal) activities of mind. But skeptics assume it refers to illusion, derangement and a variety of non- normal or abnormal clinical psychopathologies."

> R E M O T E - V I E W I N G - One Of The Superpowers Of The Human Bio-Mind;.
> *Remote Viewing and its Conceptual Nomenclature Problems* by Ingo Swann (09Jan96)

"We tried a lot of things. Like I always tell everyone, we "improved" on the Ingo (Swann) method a thousand times in a thousand ways. But our bottom line always had to be accuracy, so we had to keep track of the improvements. Most of those times, the resulting data showed that the end result of our "improvements" was to have the accuracy drop down and down and down. Those things which proved over time to work, we kept. Ingo will be the first to tell you that what we did and taught to new people coming into the unit wasn't his "pure" method. The minute someone would come back from Ingo's training, we would try to see if there were some way to make what they had learned work better in a military/political/espionage setting. Some things did work, and they are now incorporated into the "military" method which passes for the Ingo Swann method."

> E-mail from former Operation Star Gate
> Data Base Manager Lyn Buchanan

Our Introduction To Remote Viewing

An Review of the terminology, history and capabilities targeting

The Former Princess Of Wales Diana Spencer's 1997 Auto Accident

(For the cover story, see the Table of Contents)

"*When they* (University researchers) *did produce an incredibly accurate response during an experiment, it was in even a moderate sense "unnerving." In a greater sense, it was "earth shattering." As* (Stanford PhD) *Russell* (Targ) *implied, for some it was even "terrifying". In no case, was it ever taken lightly, as it always had a tendency to alter one's perspective towards reality and/or our place within it.*"

~Medal Recipient Joseph W. McMoneagle~

Princess Diana Introduction

It was the peak of the Twentieth Century's Cold War [1945-1990]. The United States, the old Soviet Union, and the People's Republic of China were striving to find new ways to get an intelligence edge. During the years 1968 to 1972, the United States obtained reports that scientists in the Soviet Union had had some success with a telekinesis program that introduced atrial fibulation into frog hearts causing a heart attack. Realizing the program could target key military and political leaders, and so driven by a threat assessment, the Central Intelligence Agency funded a Stanford University think tank in Menlo Park, California - the Stanford Research Institute (SRI) - to conduct an analysis about what humanity through the ages has pondered.

The doctors were to determine scientifically if psy-functioning could be taught, quantified and directed within written protocols. If so, did this represent a credible threat to the people of the United States? Their highly classified "Black Ops" program lasted from early 1972 until November 1995.

Under the most extensive and stringent experimentation that two PhD's could devise, the SRI, supported by other labs and the Army, developed mankind's first "psychic" protocols. "This led to greater understanding of everything from methods of evaluation, to establishing statistical standards, to how a human brain might be appropriately studied."[i] When their findings were made public, many in the academic community were privately stunned.

Eventually this covert military effort focused on real world data collection. As the years of research, analysis and application moved through the 1970's and 80's, Army brass with wholly personal motives, would attempt to quash the program even when research costs did not impact their budget. "All the funding had been approved on a year-to-year basis, and only then based on how effective the unit was in supporting the tasking agencies. These reviews were made semi-annually at the Senate and House select subcommittee level, where the work results were reviewed within the context in which it was happening."[ii]

Fortunately, for The People, the program was given different code names and moved around various Defense budgets until much of the research and development was completed. What emerged was an incredibly "robust" database - and a process

Introduction

referred to as Controlled Remote Viewing [CRV].[1]

Much of the work took place within the 902nd United States Army Intelligence Group at Fort Meade, Maryland, whose barracks have been demolished. However, from the fastidiously maintained database emerged statistically advanced practitioners; world class viewers whose RV data was the "best in the business." Among these, one remote viewer was the first in history to be decorated with the Army's Legion of Merit and Meritorious Service Awards (with five Oak Leaf Clusters) for having made key contributions to the U.S. Intelligence community. This same individual was tasked to unlock the mysteries in this Evidential Details Mystery Series.

Obviously, accuracy is the name of the game. As with any horizon application process, purposefully moving the human brain into what is likely the mechanics of a sub-quark quantum entanglement required new terminology. As the CRV process was tested, protocols written and cautiously modified, scientists documented mental hazards to viewer accuracy. These hindrances were cataloged and their characteristics differentiated. Year after year laboratory research determined accurate mental representations could be inhibited in a variety of ways. Some of these mental distracters included:

Physical Inclemency - Knowledge of an expected disruption like a phone call or someone about to arrive during a remote viewing session.

Advanced Visuals - A fleeting thought you cannot get rid of before a session.

Emotional Distracters or Attractors - An image you do or do not want to view regardless of the tasking.

Front Loading - Knowledge of what the target is before the viewing session. If localized, it can be used in targeting a feature within the whole picture, perhaps a house in a meadow in front of a mountain. However, without neutral wording like "The target is man-made" the object is generally rendered unworkable.

Analytic Overlay [AOL] - If a viewer is not informed about the target

---

[1] This may also stand for Coordinate Remote Viewing when longitudinal and latitudinal target coordinates are used.

Princess Diana Introduction

and not front loaded but still has personal information about it, that knowledge may pollute the information stream rendering the session unworkable. Analytic Overlay can be a problem for any viewer. According to the military's former #1 remote viewer:

Joseph McMoneagle - Analytic overlay - CRV [Controlled Remote Viewing], **as a format or method for learning remote viewing, offers a structure within which you can discard or identify specific elements within a session for which you are certain or not certain. Analytic Over-Lay (AOL) being a common label for something that falls within the "uncertain" category. However, when studied (under laboratory conditions), there is evidence that fifty percent of the time, information labeled as AOL actuality, wasn't.**

I have observed just as many times, someone being smacked up against the side of the head while attempting CRV because they had strayed from the given format and slipped into AOL. I think that sometimes you may forget that CRV was developed within the hallowed halls of SRI and was taught there for years. I saw very little difference in the AOL pitfalls with CRV and other methodologies. I did see that to some extent it was a highly polished technique, which was more easily transferred through training.

With this quick overview of the subconscious transference of recollections, we turn to the remote viewing of the Princess Diana Spencer's accident in the early morning hours of August 31, 1997. As this researcher found, how one targets is critical to the result. In the fall of 1997, the massive press coverage of Princess Diana's accident and funeral emerged as a very real overlay problem. The Hotel Ritz in Paris, France rather than the crash site was targeted. There had been much less news coverage at the hotel. At the time, this target was less than two months old. No accident report had been completed. An envelope, with a second target envelope inside, had been mailed to Joseph McMoneagle's home with nothing more than the targeting coordinates and a date. A skeptical *Life Magazine* reporter was on hand as an observer to write a story.

Mr. McMoneagle requested I submit a target. The viewing event started at 11:49 am on October 29, 1997. What makes these sessions interesting is that the reader can sense the Intelligence

Introduction

intellect. Having viewed 1200 targets in just the last two years of the military's Operation Star Gate alone, this job would reasonably have been assigned to the only viewer to participate in the program for twenty-three years. What was submitted was:

Target Envelope No. 102997 - (no additional information other than what's sealed within the envelope.)

* * *

As her size nine shoes hit the airport tarmac the former Princess of Wales Diana Spencer, 36, knew she was entitled to an escort by that special branch of the French Interior Ministry charged with guarding visiting dignitaries - the Service de Protection des Hautes Personalities (SPHP). But there would be no need of the service once she left the airport. This was to be a private visit.

Diana was returning from a yachting vacation in the Mediterranean off Northeast Sardinia. She and Emad "Dodi" Al-Fayed, [1955-1997] had been aboard the Fayed family's $27 million dollar (US$39.5m/2015), 195 foot yacht *Jonikal,* with 16 its crew members.

At this point, "…in her relationship with Dodi Fayed she was displaying a new facet. In some ways a late developer, she had grown up and was simply having some adult fun."[iii] But the couple had been stalked by high-speed paparazzi boats wherever they went. On their last afternoon, they came ashore at the Cala de Volpe in Sardinia and the, "Paparazzi swarmed around them like bees, flashing away."[iv] Forced back to the boat, "Things came to a head when a scuffle broke out between three paparazzi and several members of the *Jonikal*'s crew."[v]

At about the same time, hundreds of miles away, a 73 year-old grandfather, Edward Williams, walked into the police station in Mountain Ash, Mid Glamorgan, Wales. He reported to the police he had had a premonition Princess Diana was going to die. The police log, time stamped 14:12 hours on August 27, 1997, stated:

"*He* [Williams] *said he was a psychic and predicted that Princess Diana was going to die. In previous years he has predicted that the Pope and Ronald Reagan were going to be the victims of assassina-*

Princess Diana Introduction

tion. On both occasions he was proven to be correct. Mr. Williams appeared to be quite normal."[vi]

Based on his previous record the police passed this report along to the department's Special Branch Investigative Unit.

Fed up with the non-stop press hassle, on Saturday August 30, Dodi and Diana boarded the Fayed's Gulfstream IV jet at Olbia airport in Sardinia and flew north. They arrived at Le Bourget Airport about 10 miles north of Paris, France at 3:20 p.m. Fayed's butler Rene Delorm recalled, "Unfortunately, we had a welcoming committee of about ten paparazzi waiting for us."[vii] About 600 feet (183 meters) away was a Mercedes and a Range Rover. "We had all seen the paparazzi, so we moved quickly. We wanted to get out of the plane and into the cars as fast as possible. (Body Guard) Trevor (Rees-Jones) was the first out of the jet..."[viii]

The entourage had a police escort from the airport up to France's highway A-1 leading to Paris. But as they entered the expressway, reporter's cars and two man motorcycle teams immediately dogged them. The paparazzi were armed with powerful, maximum strength, flashes to penetrate deep into the car. Philippe Dourneau, 35, was Dodi's chauffeur. But in the Range Rover vehicle there had been a switch. Assistant Chief of Hotel Security Henri Paul was at the wheel. It is unclear why Paul was chauffeuring and not at the Ritz Hotel as acting Security Chief.

Once on the highway, Dodi instructed Dourneau to pick up speed in an attempt to elude photographers. What ensued was a high-speed pursuit with motorcycle cameramen weaving in and out shooting pictures. The motorcycle whirl was so intense Diana reportedly cried out in alarm that someone could get killed.[ix]

"Then a black car sped ahead of us and ducked in front of the Mercedes, braking and making us slow down so the paparazzi on motorcycles could get more pictures. They were risking their lives and ours, just to get a shot of Dodi and Diana riding in a car. "*Unbelievable*", exclaimed butler Rene Delorm.[x]

Dodi was not accustomed to this and after their high seas harassment, his patience was running thin. Pursuing for miles, the paparazzi then used phones to notify photographers ahead to form another gauntlet on the next highway segment. The Fayed cars split

Introduction

up in an attempt to divide the photographers. Some pursued Henri Paul as he drove to Dodi's apartment to deliver the luggage. Finally, the Mercedes made it to Bois de Boulogne on the outskirts of Paris to visit the Fayed's Windsor Villa. They arrived about 3:45 p.m. Then they were off to the Ritz Hotel in downtown Paris at 4:35. Alerted by the cameramen the hotel entrance was packed with photographers which in turn generated curiosity seekers in the general public.

Once inside the hotel, Diana checked into the second floor Imperial Suite and went to have her hair done. She also made some phone calls. After the accident, London's *Daily Mail* correspondent Richard Kay stated that Diana had called him saying she was going to complete her contractual obligations through November and then go into private life.

Another call was made to psychic Rita Rogers whom Diana had been in contact with since 1994. Just three weeks earlier, on August 12, Dodi and Di had visited Rogers for a reading on Dodi. She warned him not to go driving in Paris. "*I saw a tunnel, motorcycles, there was this tremendous sense of speed.*"[xi] Uneasy, Rogers reminded Diana about her readout concerning a Parisian tunnel saying, "*...remember what I told Dodi.*"[xii]

At seven o'clock, they left the hotel for Dodi's apartment at Rue 1 Arsene-Houssaye arriving at 7:15 p.m. Here the couple found the street so crowded they could not even open the car door. "The paparazzi literally mobbed the couple," said (32 year old former Royal Marine Kes) Wingfield. "They really disturbed and frightened the Princess, even though she was used to this. These paparazzi were shouting, which made them even more frightening. I had to push them back physically.'"[xiii]

From their third floor apartment, butler Rene recalled:

"*...I could see they were being mobbed. I heard the shouting, saw the flashes going off and watched a security guard shove one of the photographers. Dodi did his best to shield Diana as Trevor and Kes fought to clear a path to the door...The princess was ashen and trembling, and Dodi was angry as they stalked through the apartment door...*"[xiv]

This was the way it was going to be. Rumors were rife about

Princess Diana Introduction

a marriage proposal and some wealthy publishers made it clear big money was available to the photographer that got the "million dollar shot". But no million dollars had been budgeted.

Later, after things settled down and Dodi had returned from shopping for two rings at the Repossi Jewelry Boutique, Rene recounted, "I met Dodi as he walked through the kitchen doorway, his eyes gleaming with excitement. It was then that he showed me the ring.[2] *'Make sure we have champagne on ice when we come back from dinner,'* he told me urgently. *'I'm going to propose to her tonight!'*"[xv] Elated, he also phoned this proposal news to his cousin Hassan Yassin that evening.[xvi]

Dodi had the Hotel staff book a 9:45 p.m. dinner reservation at the fashionable restaurant Chez Benoit on the Rue Saint Martin. He also phoned the Ritz staff he would not be returning. As a result, Security Chief Henri Paul departed for the weekend at 7:05 p.m.

At 9:30 p.m., Dodi and Diana left the apartment for dinner but could not get through the crowd at the restaurant entrance. It was clear they could not enter a restaurant together. The enormous number of paparazzi forced Dodi to cancel their night out. The Press was controlling his special night with his special lady. A frustrated Dodi decided they should make the four mile drive to the Hotel Ritz where they could dine in France's only "safe" restaurant. But Security Chief Henri Paul had gone for the weekend and the abrupt change left the hotel staff with no time to prepare for their arrival.

When they arrived at the Ritz, another press riot broke out. It took Diana two whole minutes to negotiate the camera gauntlet the 20 feet from the front door drive-up to the hotel turnstile. The security camera time stamped her entrance at 9:53 p.m. Security man Wingfield said:

"*I had to protect her physically from the paparazzi, who were coming really too close to her*[.] *Their cameras were right next to her face.*"[xvii]

Dodi was furious and started shouting at his employees about no security to shield the 10-second walk up from the driveway. Shaken, the press savvy Diana wept in the lobby. Everyone was

---

[2] Dodi received a US$100,000/month ($146.5/2015) allowance from his father.

Introduction

upset. With the owner's son angry, and the security force completely embattled, a decision was made to call the Security Chief back to work. Francois Tendil called Henri Paul's cell phone at 9:55 p.m.

Once safely in their room, Dodi called his father Mohammed Al-Fayed at approximately 10:00 p.m. He said the two would announce their engagement the next week when Diana returned from England.[xviii] "Diana always had the children for the last few days before they went back to school at the start of a new term, so that she could get everything ready and make sure they had the right kit."[xix] On Friday, she had called to confirm her boys would be at the airport to meet her on Sunday morning.

Dinner was ordered from the hotel's Imperial Suite restaurant. Diana's last meal was scrambled eggs with mushrooms and asparagus, then vegetable tempura with fillet of sole. As Di and Dodi were trying to dine normally, Henri Paul pushed his way back into the hotel through the paparazzi.

For this targeting, the Hotel Ritz Building was tasked using the proper date, time, and location coordinates. As Mr. McMoneagle looked at a double blind envelope, he started:

McMoneagle - I find myself standing next to a man who is inside some kind of a public building. He is approximately five feet, ten inches in height, good build, good condition physically. He weighs about 165 pounds, is clean shaven, light brown hair, right handed, 38-40 years of age, and is not British or American; meaning he probably has another language other than English as his native tongue.[3]

Upon his return, Henri Paul waited around the Ritz for about two hours. He allegedly had a couple drinks at the bar. The Ritz security cameras recorded his behavior which would be used for future analysis. As Chief of Security, he was certainly aware of their placement and recording capabilities.

McMoneagle - Building interior - Where he (Paul) is within the building is inside of a very elaborate corridor. It runs the full length of the building and has lots of gilded paint, mirrors, thick carpets, lots of flowers, and is very fancy. The

---

[3] Paul was 167 lbs. and he was 41 years old. He had brown hair and was also balding. His native language was French. He spoke fluent English and some German.

corridor runs straight out to a front entry which is well lit and very busy (even though my sense is that it is very late at night). There is an area off to the right of this corridor which has a lot of dark paneling and dark colors with a long bar or type of counter. So, this may be the reception area of the hotel or something like that.

Where he (Paul) **is standing is where the main corridor intersects with a short corridor that runs off at a ninety degree angle to the left. It intersects with some kind of a smaller staff or receiving area; perhaps a back door to the building. It is recessed and that is where his car is parked.**

The Etoile Limousine Company manager Jean-Francis Musa, 39, provided six luxury cars to the Ritz Hotel for their exclusive use. This Mercedes was licensed as a Grande Remise auto meaning only a licensed chauffeur was authorized to drive it. Henri Paul did not possess those credentials.

McMoneagle - Driver orientation - I believe that he (Paul) **drives a cab or limo...on the side, because I associate him with a car, which is parked outside and he is thinking about this car, or it seems to occupy his thoughts for some reason. He is mostly interested with driving from point A to point B. I believe he is not alone and get a strong feeling of mixed male/female in energy; which either means his passenger will be gay, or consist of two people--a male and a female.**

Limo is not a stretch limo but a short, black and formal kind of car. I get an impression of a Mercedes emblem or some kind of emblem like that, so I'm assuming it is a very expensive car, could be a Mercedes.[4] **It is formal and black with an extended foot space in the back seat. Four doors. It is very heavy and my sense is that it might be equipped for important passengers — e.g., bullet proof glass, armoring, hardened tires, etc.; which leads me to believe that at least one of the passengers** [Trevor Rees-Jones, 29] **might be a body-guard** [but]

---

[4] The Mercedes S 280 sedan, valued at about $100,000 (US$146.520/2015) was engineered with eight advanced safety systems. The car had a reinforced chassis and roof. It had energy absorbing front and rear end crumple zones with electronic traction control. It also had an electronic ESP sensing system, which monitored trajectory with wheel speed to sense cornering speeds.

Introduction

this may be Analytic overlay caused by the excessive feelings of security surrounding this vehicle and driver.

* * *

Information about Henri Paul's mixed motivations have come to light in the years since the accident. Born one of five brothers on July 3, 1956 in the port town of Lorient, France, he had a Bachelors Degree in Mathematics and Science from the Lycee St. Louis and had won several contests for his skill as a classical pianist. He became a pilot in 1976 but was unable to qualify as a jet fighter pilot when he joined the French Air Force in 1979. Paul did however achieve the rank of Lieutenant while assigned to Security in the French Air Force Reserves.

In 1986, Paul helped setup Ritz Security. He went on to become Assistant Director. On the day of the accident, he was carrying 12,560 francs (US $2,280) and his savings account passbook.[5] Where the money came from is unknown, but he was one of only two men in France that had access to the automobile conversations of Dodi and Di. The ability to advise the press of their plans would have been of great value.

Personal adversity. Henry Paul had recently been passed over for promotion a second time by Hotel Ritz management. The first disappointment had come on Jan 1, 1993 when the nod went to colleague Jean Hocquet even though Paul was obviously in position as the number two security man. Now again, effective June 30, 1997, as "Deputy Chief" he became the defacto head of a twenty person security team while Ritz Management searched for another chief. Now vulnerable, Paul had been informed of this exactly one month before the accident.

Post mortem tests stated Paul had consumed two antidepressants called Fluoxetine and Tiapride before the accident. Fluoxetine is the active ingredient in Prozac and together these drugs are commonly used to fight alcoholism. When alcohol is

---

[5] Henri Paul may have charged the equivalent of US$2,250 (1997) per surveillance event and simply had an additional $30 pocket money that day. His salary was reported at $40,000 ($58,600/2015) per year.

introduced, the intoxicant effect is multiplied. On September 17, a more sophisticated laboratory's final report was issued. It stated that Henri Paul had been in, *"moderate chronic alcoholism for a minimum of one week."*[xx] Once this became public, the Ritz's attorneys and Mohammed Al-Fayed found themselves on the defensive. An unlicensed employee now appeared criminally negligent in a multiple wrongful death accident while in Hotel Ritz employment. It became the million dollar shot vs. the Al-Fayeds.

The intoxication driving limit in France is 0.50 grams per liter. One lab report stated Henri Paul's blood alcohol level was 1.87 g/l. This is the equivalent drinking time for eight or nine shots of whiskey in what was found to be an empty stomach. A second, private laboratory's more moderate findings were used in the Final Report. The Paris Prosecutor's Office Report stated:

"On this particular point, numerous expert's reports examined following the autopsy on the body of Henri Paul rapidly showed the presence of a level of pure alcohol per litre of blood of between 1.73 and 1.75 grams, which is far superior, in all cases, than the legal level.

Similarly, these analyses revealed as [did] *those carried out on samples of the hair and bone marrow of the deceased, that he regularly consumed Prozac and Tiapridal, both medicines which are not recommended for drivers, as they provoke a change in the ability to be vigilant, particularly when they are taken in combination with alcohol."*[xxi]

So had Henri Paul been out drinking? It is known he returned to the Ritz two hours and fifty minutes after departing. But no one knew where he was when he received the Ritz phone call. Investigations into who had seen Paul failed to provide a single witness. In Paris, in the fall of 1997, there was a real fear of liability for anyone acknowledging Paul had been drinking in their establishment. Nonetheless, the French media reported "*someone*" saw Paul drinking "aperitifs" between 7:05 and 10:08 p.m. that evening. "Someone" is wide open. It means that after he got the call to return at 9:55 p.m., he dallied almost another quarter hour before departing which is hard to believe given the tone of the call. Until now, the critical question about where and what Paul was doing before returning to the hotel remained unknown.

Introduction

McMoneagle - I think he was in fact sitting in a small restaurant or coffee shop, very near where he lives. Maybe even on the corner near his house. He was alone as far as I can tell. I think he was in fact drinking coffee. I do not think he was depressed, at least not more than usual. Also, regardless of what might be said, I **DID NOT** get a sense that he was drunk. It is remotely possible that he was taking some kind of a medication but I doubt it.

Coffee! Not drunk! This flew in the face of the formal investigation. We were now privately aware, months before the controversy started, Henri Paul was not drunk.

Henri Paul was a pilot. Research indicated it was impossible to reconcile allegations of alcoholism with Paul's recent physical examination. Unbeknownst to the authorities issuing the report, just two days before the accident, Paul had completed a "rigorous" physical examination to renew his pilot's license. His *Certificat D'Aptitude Physique et Mentale* showed, "No signs of alcoholism."[xxii] A direct medical conflict supporting McMoneagle. Was Paul really fighting alcoholism? Six months after these sessions, the Ritz Hotel security videos further reaffirmed our data.

Behavioral Psychologist Dr. Martin Skinner commented in Fulcrum Productions documentary for ITV. The doctor stated there were no behavioral signs of drunkenness as Henri Paul waited for Dodi and Diana.

Skinner: *I don't think there is evidence, from the video, that can suggest he looked drunk. The pictures of him walking up and down the corridor are straight and smooth. He is standing very still and there is nothing in his demeanor, from these videos, to suggest that there are any problems with his competence in this situation.*[xxiii]

Next came a statement from Trevor Rees-Jones, the front seat bodyguard sitting next to Paul. About intoxication, he said:

Rees-Jones: *I had no reason to suspect he was drunk. He did not look or sound like he had been drinking. He just seemed his normal self. He was working. He was competent. End of story. I can state quite categorically that he was not a hopeless drunk as some have tried to suggest. I like to think I have enough intelligence to see if the guy was plastered or not – and he wasn't.*[xxiv]

Princess Diana Introduction

Session Sketch

This drawing provides a rare glimpse intelligence level RV artwork. In this exercise, people and not the building were targeted. But, this sketch could be the third floor at the North Korean Embassy in Beijing, China, or any building, anywhere, anytime. As a person was the target, the Hotel Ritz Paris first floor was roughed out at midnight on August 31, 1997. Points of interest are:

1) At the top of the page, the words **Big Bldg** appear;
2) The various circles with an **X** inside indicates where people were standing at approximately 12:15 a.m. on August 31, 1997.
3) On the left, the **Main Door** is shown with an **X** representing the doorman. As the hall extends to the right, the various rooms are notated.
4) Toward the bottom is a **Business** area. As you walk from the front door, **"There is an area off to the right of this corridor which has a lot of dark paneling and dark colors with a long bar or type of counter."**
5) At the top is an **Alcove** with two people inside. These individual's backgrounds – conversations – futures – mental states - deaths can be targeted at any time in the future.
6) Where the hallway comes to a junction there is a **Man**. This is Henri Paul as he monitors the activities in both corridors. What were Paul's private thoughts? **"I associate him with a car which is parked outside and he is thinking about this car, or it seems to occupy his thoughts for some reason."**
7) Behind Henri Paul is the **Laborer Area**. Next to this is the drawing date and time documenting who was where when.
8) The hallway to the **Side Door**, **"...intersects with some kind of a smaller staff or receiving area; perhaps a back door to the building. It is recessed and that is where his car is parked."** That recessed area is shown.
9) McMoneagle also shows the **Formal Black Limo**'s position by the back door and correctly identified the automobile's color and manufacturer's hood ornament (bottom right).

Introduction

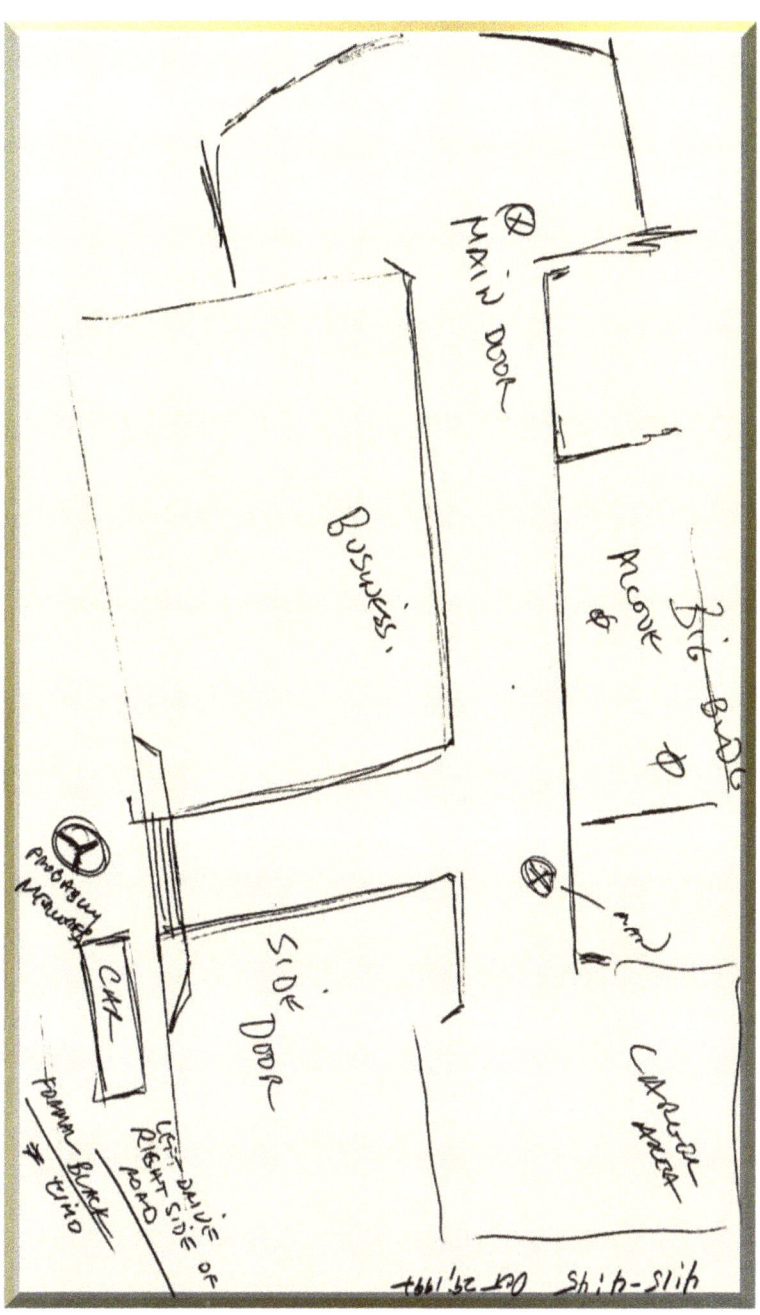

McMoneagle RV Art – Evidential Details ©1997
Hotel Ritz Paris first floor with car (lower right) as viewed from Virginia.

Princess Diana Introduction

Neither the bodyguards, nor Dodi, or anyone else at the Hotel detected anything unusual in Paul's behavior. But there was more.

Paul's blood was next reported as containing abnormally high carbon monoxide levels - twenty percent too much. How this happened has never been determined. But doctors agree it is impossible for a forty year old man, with that much poison in his blood stream, not to look and feel sick - too sick for high speed urban driving. When the press advanced the idea car exhaust was the source of Paul's poisoning, Dodi's father, Mohammed Al-Fayed, put the obvious question: *"How did Henri Paul get 20% carbon monoxide in his blood when my son had none?"*[xxv]

The obvious question is how you can get that much CO^2 into someone's blood stream when, due to an instantaneous death there was no breathing, and the engine had stopped.

During his last month Henri Paul had come to know what it was like to assume the Security Chief's responsibilities while the Ritz Hotel interviewed. He must have been concerned an outside hire may not be as accommodating as his previous colleague boss had been. After setting up the Ritz security operation, and with a decade of service, Henri Paul now faced the possibility of being forced out by a new supervisor uneasy about his hotel security experience. Clearly, Ritz management was not taking care of Paul as a career professional.[6]

Another component of the Henri Paul enigma concerned the fact that most nations have an Embassy in Paris and many dignitaries and diplomats stay at the Ritz. Stories started to appear that Paul was in the employ of various "foreign and domestic" intelligence services. Then it was discovered he had one million francs [US$200,000–$250,000] spread between eighteen bank accounts in an attempt to disguise the fact. Al-Fayed would later make the claim Paul had spent at least three years working for British intelligence. Where he got this information, or if it is true, is unknown. Paul was also allegedly in contact with the Direction General de la Securite Exterieure [DGSE] - French Intelligence. So,

---

[6] The Hotel Ritz subsequently hired a former Scotland Yard Chief Superintendent John MacNamara. His background in criminal intelligence management and investigations was substantially different than Paul's Air Force Reserve security credentials.

Introduction

we were left with a feigned alcoholic dead man; with employment and big money surveillance concerns; ordered to violate multiple traffic laws; by a romantically aggravated boss in love with the world's foremost beautiful woman.

Henri Paul was uncertain about his future. He had to have been anxious about protecting his access to the Hotel Ritz time and date stamped video-monitoring system. He must have been concerned about his ability to generate good income by documenting high profile business people or foreign dignitary's arrivals and departures.

But all of a sudden, that night there was a positive side to the whole discordant affair. A rare opportunity to make a positive impression on the owner's son was at hand. In the wee hours of August 31, 1997, it would have been impossible for any driver to presume to caution a provoked Dodi Al-Fayed about safe driving on nearly deserted streets. As characterized by French Union Official Claude Luc:

> *"If one of the Fayeds gives you an order, you follow it. No questions asked."*[xxvi]

Whatever his prospects, Security Chief Henri Paul was illegally behind the wheel again. He was laid to rest in Lorient, France on September 20, 1997. Father Léon Théraud gave the sermon at Sainte Therese Church.

* * *

On Saturday night, now Sunday morning August 31, a physically aggressive horde of stalkarazzi and other onlookers, estimated at approximately 130 people, jockeyed for position at the front door of the Hotel Ritz Paris. Diana Frances Spencer and her boyfriend Dodi, son of Egyptian born multi-millionaire Mohammad Al-Fayed, needed a second car to exit the hotel's back entrance. Because of the paparazzi, a front door - back door scheme had been set-up for their return to Dodi's apartment. Dodi would take Diana out the back leaving his personal Range Rover in front as a decoy.

McMoneagle – Car is parked on the right side of the road (right side driving) which would rule out England,

Princess Diana Introduction

Bahamas, Hong Kong, Japan, etc. It is night and it is dark. The time for this event is current, probably 1985 to 1997. I will try and bring that down to a shorter period later.

The tag on the limo is elongated, with letters and numbers--which is a European style of tag (License 688 LTV 75). My sense is that there may actually be two colors of tags on this car, or that it has inter-changeable tags, which are changed, dependent upon where it is being operated. One is yellow with black lettering; the other is white with black lettering. It may be that there are two different colored tags on the car simultaneously—one color on one end, one color on the other.

This is a superb surveillance example. The yellow license with black lettering was on the rear bumper. As it turned out, the color license designated a private car. The white tag is a "for hire" vehicle. From this the reader can gather the type of information available through remote viewing should this car have been driving a foreign dignitary.[7]

After some hallway discussion, Ritz chauffeur's Philippe Dourneau and Jean-Francois Musa drove two decoy vehicles to the hotel's front door. The night was clear. The temperature was 77 degrees [25C]. Their engines were revved up as Dodi and Diana hurried out the back door at 12:20 a.m.

Diana's last few minutes on earth were now inexorably caught-up in the emotional web of her incensed boyfriend and his driver's employment needs. Some paparazzi across the Rue Cabman observed them as Trevor, Diana, and then Dodi came through the turnstile and got into the Mercedes. Henri Paul pulled out and the chase was on.

McMoneagle - Believe the car is the main focus of this target. The man [Paul] may also be of interest.... I believe this target has to do with an accident that probably occurred either in the very late night hours or possibly very early morning hours. Traffic is very light and the streets are very quiet. Get a

---

[7] In Foreign Relations, these plates could indicate a restricted territory vehicle. If unauthorized, remote viewers could be tasked on who issued both types of plates to the same party. This inquiry would remain secret, while perhaps unmasking a corrupt government official, or a mole in the host country's bureaucracy.

Introduction

sense that there are few cars about, in a place which is usually crawling with cars.

The Mercedes is moving very fast from what apparently is a northwest...direction. Have a sense that it goes over an overpass or cloverleaf kind of interchange which then drops straight down into a tunnel.

Associated Press
A back door security camera photograph time stamped 12:19 a.m. just before they departed. It shows Henri Paul (left) conversing with Dodi and Diana with Trevor Rees-Jones in the background above.

The car traveled toward the Seine River's westbound express street referred to as the Cours la Reine. Then they entered the Alexander III & Invalides Tunnel Bridge. The tunnel is 330 meters (361 yards) long.

McMoneagle - It [Mercedes] **then exits the tunnel and covers a large curve of open road which enters another tunnel like area, only this second tunnel is not enclosed completely. Have a sense of concrete tiers on one side... Vehicle is moving very quickly, perhaps in the neighborhood of approximately 100 MPH** [162 km/h], **maybe even a bit faster (in some spurts or straightaways).**[8]

---

[8] The curve in the road is 480 meters [.3 miles] in front of the next tunnel, which provides an acceleration area. But with a subsequent curve and dip, it was not possible to negotiate that section of highway at high speed.

Princess Diana Introduction

In my opinion, the driver was driving way beyond the speeds that would have been comfortable for the place and time. I believe he was well trained as a driver but not for the place or speed at which he was driving. I have a sense the driver was doing his damnedest to carry out the instructions of those he was carrying, but was operating at speeds and conditions that even he was never really trained to drive within. I think he was the professional here and was being egged on by the passengers.[9]

These sessions took place approximately ninety days before the release of the official fifty-two page report entitled, *Accident de Passage Souterrain de l'Alma. Paris Dimanche 31 Aout 1997, Oh25. Propostition d'Analyse Scientific et Technique. Synthese et Conclusions.* French Engineer Jean Pietri had been commissioned to write an engineering crash analysis, which went on to verify this earlier remote viewing material.

The distance from the first tunnel to the Pont de l'Alma tunnel is 1.2 kilometers (.75 mile). The speed limit is 30 mph (48km). It is here that published accounts differ. Apparently, three people witnessed four to six paparazzi motorcycles attempting to pull alongside the speeding Mercedes. Other accounts say the paparazzi were a quarter of a mile behind when the Mercedes entered the tunnel. In either event, it was all futile. Notified by telephone, reporters had already assembled at Dodi's apartment entrance, million-dollar picture in mind.

McMoneagle – The Mercedes pulls out to pass a slower moving vehicle at a point in the road where the road ahead rises upward to a secondary overpass. Because of the rise in the road, the driver can't see on-coming traffic in time to avoid it, specifically at this speed.

The final report showed this was correct. French accident investigator Jean Pietri subsequently stated:

"To our surprise, we observed that the field of view is extremely limited. Passing cars disappear from sight well before they actually enter the tunnel because the descending road is obscured by a

---

[9] McMoneagle was correct on this detail. Dodi knew Paul had attended special driving courses in Stuttgart, Germany from 1988 through 1993, receiving high marks.

Introduction

retaining wall. To the left the field of vision is blocked by a row of trees."[xxvii]

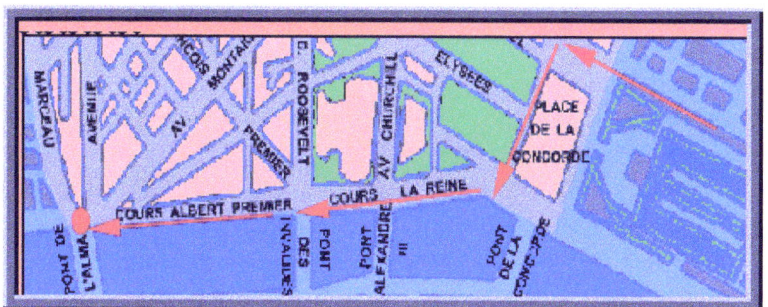

A view of their route along the Seine River. The red arrow (top right corner) points the direct route to Fayed's apartment.

About 40 meters (44 yards) in front of the tunnel the Mercedes hit a gap in the pavement, which further destabilized control. As the car passed a white Fiat Uno at break neck speed Henri Paul saw another car dead ahead.

McMoneagle – I believe he sees an on-coming car which appears to be some kind of a black or dark green sedan. I want to say Citroen, but I'm really not sure. Probably a smaller two door car, two passengers; get a sense of dark green or green-black combination, which could mean a green car (body) **and black** (trim).[10]

Mohammad Medjahdi was driving a Citroen BX with his girlfriend Souad in the tunnel ahead of the Fiat Uno.

McMoneagle – Dodi's last words - Have a fleeting sense that he [Paul] is being ordered to go faster and to do more erratic things, to avoid something. He is essentially being ordered to do what he is doing.

To avoid the on-coming traffic, the Mercedes driver swerves hard to the right and catches the small car he is passing [Fiat Uno] with his rear bumper. Car that was passed was hit. As a result, the Mercedes slews around left, just misses the on-coming car, which [it] **has just passed, and the driver then begins to over-correct his steering.**

---

[10] McMoneagle was obviously in the car looking through the Mercedes windshield. The use of "oncoming" describes the overtaking of cars. It does not refer to opposite direction traffic flow.

Princess Diana Introduction

Months after these sessions, French engineers confirmed the Mercedes did nick the Fiat Uno and over corrected to the right. Some tail light/head light debris was found.[11] Engineers estimated that if the Mercedes was going 100 miles per hour the debris would have rolled sixteen meters (52.5 feet). That hit took place outside the tunnel and it is here the 18.9m (62ft) tire skid mark begins.

McMoneagle - The Mercedes hits the side to left slews across and hits the right, then swings back to the left, where it catches what appears to be a concrete tier or pier (#13 pillar) **of some kind, concrete pilasters, or some kind of upright** (steel reinforced) **concrete dividers, which it hits nearly head on.**

At 12:24 a.m., there was an explosion sound in the tunnel. The subsequent engineering report confirmed Henri Paul's last evasive actions was viewed correctly. Various eyewitnesses recounted the collision. "Gaelle L., 40, a production assistant stated:

"At that moment, in the opposite lane, we saw a large car approaching at high speed. This car swerved to the left, then went back to the right and crashed into the wall with its horn blaring. I should note that in front of this car, there was another, smaller car."[xxviii]

McMoneagle - The Mercedes apparently nearly goes end over end rear to front, but doesn't quite make it [over the top], **instead spinning twice and winds up pointing back in the direction it was coming from.**

The car spinning 1 1/2 times remains unconfirmed. But there was enough inertia for the car to have spun 540 degrees when the rear wheels were off the ground. The impact was so hard that the forward roof area was crushed down to the level of the driver's knees. This is further substantiated by the fact Diana was found facing backward in the back seat, which would not have happened with a simple 180-degree turn. N*ewsweek Magazine* reported French police estimated the car had slowed down to 85 mph at the point of impact.[xxix]

The entire trip had taken about four minutes. Trevor Rees-Jones could only recall the Fiat Uno.

---

[11] The tail light pieces found in the tunnel belonged to a Fiat Uno manufactured between May 1983, and September 1989 by Seima Italiana. The white paint chips were called Bianco Corfu. When found, the car had been repainted.

Introduction

Rees-Jones: *"It seems to me there was one white car with a boot which opened at the back* [hatch back]*, and three doors but I don't remember anything else."*[xxx]

He did not leave the hospital until October 4 - thirty-four days later.

Aware Henri Paul did not have alcohol in his system we sought clarification to research about drugs in his blood stream.

McMoneagle - Substance review - I believe if the driver had drugs in his system, whatever kind they were, they were not there by his own hand. I have this sort of strange feeling that he was not deliberately drugged to hurt anyone, but maybe he was drugged to get the car stopped along the route for the "photographers" to get their shots. In other words, his control was tampered with by outside influences. I don't think he was drunk, possibly drugged, but not drunk.

Here the research came full circle. The paparazzi had attempted to slow the Hotel Ritz airport shuttle vehicles earlier that afternoon on the drive from the airport. Once it was discovered, Henri Paul had been an informant for domestic as well as foreign intelligence services we went back to McMoneagle. Could the British government have been involved?

McMoneagle - My sense is that MI-5 (British Intelligence) **did not put the stuff in his drink. However, one might contemplate that if he** [Paul] **was willing to take money from foreign intelligence operatives, he most certainly would have been open to taking money from the Paparazzi. Maybe they were hedging their bets by having a small "drink" with him in the bar before he started driving.**

And what of the high carbon dioxide levels in Dodi's blood stream? Since this viewing, there were reports of a carbon monoxide suicide in Paris that night.

McMoneagle - You have to open your perception a little bit here. He did not have to have any evidence of CO2 in his blood for them to find CO2 in a blood sample. You only have to switch the samples at the hospital, the morgue, or the lab. Or, pay off the guy who is doing the tests. You could also conceivably rig the test equipment. Also, there are drugs, which will give a false reading as well.

Princess Diana Introduction

His being drugged enough to cause the accident could be attributed to a drug delivered in coffee, tea, or a drink beforehand. It could also have been sprayed on the inner edge of his door handle (driver's side), painted on the steering wheel, or inside a pair of driving gloves. He could have been shot with a needle delivery system, or pricked his hand, finger, leg, or almost any part of his anatomy on a delivery system getting into or out of the car. It can even be filmed across the pages of a book or map that he might have used to check directions on.

If he had a normal medical condition, they could have used a drug, which reacts violently with the drugs he is already taking for the medical condition. In which case they would either get false readings, or evidence of his medicinal drug, plus some other known drug which would not have been viewed as culprit in the event, simply because no one recognized the possible expected reaction. You also have problems with drugs which are binary in nature and can be delivered in two sittings, so to speak, where the victim gets part A in the morning with breakfast, part B in the evening with dinner, both of which are enzymes and when mixed... cause everything from hallucinogenic behavior, to strokes.

Now we turned to what Dodi and Diana where thinking.

McMoneagle - Back seat travelers - MAJOR PROBLEM: When I try to access others who might have been in the car, I get heavy [analytic] overlay and interference as relates to Diana's death in France. My head fills up with all kinds of motorcycles, and all kinds of news... that was being broadcast about the incident. I believe there were at least two others in this target car, but digging anything out of the overlay is completely impossible.

There is a sense from the people in the back seat that they want to be alone together, but again, I then get overwhelmed with all the Princess Diana stuff... and it all runs together. So, I can't begin to tell where [the] overlay begins and real data ends. Would prefer to say nothing.

It's rather interesting. I actually have not opened the envelope nor have a clue as to the real target here; but I am being overwhelmed with overlay which is self-generated. Must

Introduction

have been a lot of energy around the Princess Diana stuff. Better to just go no further with it. End of Session.

An abrupt stop, on a then well-known topic, due to analytic overlay. This is a graphic demonstration of the differences between military remote viewers, storefront psychics or hot lines. The media had been saturated with Princess Diana coverage in the period between the accident and this tasking. A psychic hot-liner would have been able to talk and bill without end about what they "saw". One Operation Star Gate military remote viewer commented, "There are many "*psychics*" who have taken this type of gibberish to a finely honed skill."[xxxi] But, when McMoneagle got to the Mercedes back seat, he stopped the session. In intelligence work when you are not sure of your viewing, you must say so. Any elaboration is unethical as in life and death situations, military viewers must stay grounded in the target's realities.

Analytic Overlay [AOL] is terminology within the Controlled Remote Viewing [CRV] protocols developed by Ingo Swann for the U.S. Military Intelligence Community at the Stanford Research Institute as they developed the nomenclature. AOL can generate bad data. So, can anything be done about it?

McMoneagle - Military research - There were a number of experiments which were run to examine whether or not a remote viewer can identify "AOL" while in session. We found that it could be rarely demonstrated. Most viewers are unable to tell (accurately or consistently) when something was AOL or when it wasn't, while in session.

Facts are; Evidence produced within labs suggests that no one methodology is capable of identifying and extinguishing AOL any better than another over the long haul.

There have been significant runs of very low AOL or displays of almost no AOL which have been done by individual remote viewers. So, there are indications that some people might have a talent for producing less AOL than others. But it does not appear to be method driven since it doesn't hold up in testing across all remote viewers using the same method.

So, why should identifying AOL be important??? It is important because, while you are attempting to learn remote viewing (regardless of method), it makes you think about how

and why you are "thinking" about something. It is meant to reduce the speed by which you automatically jump to a conclusion. It also supports the structure and keeps one within it (at least until one becomes proficient enough to no longer need it.)

After the impact, eyewitnesses saw a motorcycle 30 to 40 meters behind the Mercedes slow down to observe the accident and then accelerate away from the scene. At 12:26 a.m., the Paris Fire Department - Sapeurs-Pompiers Unit - received a cell phone call from a Gaelle who was in the tunnel. Within one minute another call went out to the "service d'aid medicale urgente" (SAMU) - a civilian emergency medical service.

Inside the wreck, Diana and bodyguard Trevor Rees-Jones were still alive. One eye witness said he heard a woman crying loudly. One of the paparazzi, Romuald Rat, indicated Diana was conscious. He claimed he told her to stay calm; that help was on the way. She remained in the car…

Aftermath

Now pandemonium broke out as the Press fought each other to get the new million dollar shot. One photographer leaned into the car to reposition Dodi's corpse for a posed picture. Someone else came with video equipment. Within five minutes, Police Officers Lion Gagliardone and Sebastien Dorzee plowed through the crowd to the car. The police report stated:

"I observe the occupants in the vehicle are in a very grave state. I immediately repeat the call for aid and request police reinforcements, being unable to contain the photographers and aid the wounded."[xxxii]

Officer Dorzee: *"I finally got to the vehicle... The rear passenger (Diana) was also alive... She seemed to be in better shape* (than Rees-Jones). *However, blood flowed from her mouth and nose. There was a deep gash on her forehead. She murmured in English, but I didn't understand what she said. Perhaps 'My God!'"*[xxxiii]

Ultimately, six paparazzi were held in connection with the

Introduction

frenzy in the tunnel. They were arrested on suspicion of involuntary homicide and failure to assist persons in danger. Excepting the 24-year-old Romuald Rat, 40 was the average age of those arrested. Twenty film rolls were confiscated providing police with the photographic evidence they needed to confirm each man's activities that night. Three paparazzi got away.

There are no Miranda rights in France, nor is there a right to call an attorney. French authorities can hold a suspect for forty-eight hours before the prisoner must be formally charged or set free. However, it is certain Henri Paul did not have to be drunk or drugged to have had an accident at that speed.

The former Princess of Wales, Diana Spencer, arrived at the Hospital de la Pitie-Salpetriere at 2:00 a.m. She was pronounced dead at 4:00 a.m. It was then she attempted to contact her son William in Scotland. "William had had a difficult night sleep and had woken many times. That morning he had known, he said, that something awful was going to happen."[xxxiv] When he was told of his mother's death he said, "*I knew something was wrong. I kept waking up all night.*"[xxxv]

At 5:00 p.m. Prince Charles, 48, flew into Villacoublay military airfield outside Paris from Aberdeen, Scotland with Diana's sisters Sarah McCorquodale and Jane Fellows. "Diana's sisters spent most of the flight to Paris in tears. The Prince was controlled but clearly very shaken."[xxxvi] By 5:40 p.m. he was greeted at the hospital by the French President and Mrs. Jacque Chirac (1995-2007). Charles was led into a room with his two ex-sisters-in-laws where Diana lay in a coffin. He asked to be alone with the body for a moment. When he came out his eyes were red. The accident was 368 days after the finalization of their divorce.

Diana's coffin, draped in the Royal Standard's yellow and maroon, was flown home by an honor guard in a British Royal Air Force BAe146 military aircraft to Northolt Air Force Base in England. She was then taken to the Chapel Royal at Saint James Place.

Undertaken by Levertons, her September 6 funeral was the largest in England since the death of former Prime Minister and Nobel Literature Prize winner Winston Churchill [1874-1965]. After the morning funeral, it was reported a million people lined the route as the body was taken from London's Westminster Abby. Different

Princess Diana Introduction

accounts estimated two to three billion people watched the day's events as the car traveled the seventy-five miles to Althorpe House. Late that afternoon her body was laid to rest on a 1,254 sq. meter (13,500sqft) island called The Oval in a lake on the Spencer's ancestral grounds. The four hundred-year-old estate was then partially turned into a tourist attraction.

On September 9, 1997, the week after Diana was buried the Al-Fayed attorney filed civil law suits against the French periodicals *France-Dimanche* and *Paris-Match*. The complaint specified invasion of privacy with willful and wanton reckless endangerment when helicoptering "stalkerazzi" got too close over the Fayed's villa in St. Tropez. But, for the Hotel Ritz, the question became who bears responsibility for the accident? Before 1997 was out, the Fayed, Spencer, Rees-Jones and Paul families had all filed papers to be made civil parties to the investigation. Under French law, this allows them to investigate the case file and participate in any damage awards. And as for the Paparazzi's fate:

"In accordance with articles 175, 176 and 177 of the Code of Penal Procedure; The examining magistrates find that there is no case to answer in the case of the state versus the above named [Photographers]."[xxxvii]

In July of 2004, after the planning, funding and construction were completed, Queen Elizabeth II personally opened the Princess of Wales Memorial Fountain in the southwest corner of London's fashionable Hyde Park.

Then, in April 2008, after a three year investigation costing $7.3 million ($8.3million/2016), a six month long British inquest report was released which included the testimony of 278 witnesses with more than 600 exhibits generating an 832 page report stating:

"Our conclusion is that, on the evidence available at this time, there was no conspiracy to <u>murder</u> any of the occupants of the car," Lord Stevens of Kirkwhelpington, who led the inquiry, told reporters as he presented his findings here. *"This was a tragic accident."*[xxxviii]

In September of 2012, the French magazine *Closer* published paparazzi photos of Diana's eldest son's wife Kate Middleton sunbathing topless while at the Queen's nephew, Lord

Introduction

Linley's French chateau. A publically released statement on behalf of the Duke and Duchess said: "*The incident is reminiscent of the worst excesses of the press and paparazzi during the life of Diana, Princess of Wales, and all the more upsetting to The Duke and Duchess for being so.*"

And as for the need to practice remote viewing protocols:

McMoneagle - Pick whatever method you intend to pursue and stick to it like glue. AOL (Analytic Overlay) **is a fact of life and this will always be so. Those of you who can eventually see your way to controlling your inner-driven or more personalized prejudice while internally processing, will probably improve somewhat in reducing AOLs, but AOLs will never entirely go away.**

CRV (Controlled Remote Viewing) **is a "method" derived from a method the military used while attempting to "train" people to understand both protocol as well as what is going on in a remote viewer's head (such as processing or the lack thereof). It was also very specifically designed to "preclude" things from being done out of ignorance (during the RV session) that might impact on/or otherwise prevent the act of successful psychic functioning from taking place; in other words, insure that RV could be replicated and would work more times than not.**

I would add that formal testing in the SRI Lab showed that regardless of technique or methodology utilized, most viewers were unable to consistently identify AOLs when asked to identify them prior to feedback. I have to say most, because "a couple viewers" were able to do so during significant runs--but this is inherently talent based and not the general or common rule. I remind you all of what is termed the "AH-HA". If it were not for the Ah-ha's, there would not have been a program. At the end of the road, almost anything is right when you have finally come to understand that it is an inherent part of our nature and then you just simply can do it.

Princess Diana Introduction

References

[i] McMoneagle, Joseph W., *Remote Viewing Secrets – A Handbook*; Hampton Roads Publishing Company, Inc. 2000 p. xv
[ii] McMoneagle, Joseph W., *The Stargate Chronicles*; Hampton Roads Publishing Company, Inc. 2002 p. 182
[iii] Simmons, Simone, *Diana – The Secret Years* with Susan Hill; Ballantine Books 1998 p.120
[iv] Delorm, Rene, *Diana & Dodi - A Love Story - By the Butler Who Saw Their Romance Blossom*, with Barry Fox and Nadine Taylor; Tallfellow Press 1998, p.144
[v] Anderson, Christopher, *The Day Diana Died*; William Morrow and Company 1998 p.114
[vi] Anderson; p.113
[vii] Delorm; p.154
[viii] ibid; p.154
[ix] The Learning Channel Presentation - *Princess Diana*; A Fulcrum Production; a Granada Presentation for ITV 1998; hereafter referred to as *TLC*
[x] Delorm; p.155
[xi] Anderson; p.99
[xii] ibid; p.166
[xiii] Sancton, Thomas and Scott MacLeod, *Death of a Princess - The Investigation*; St. Martin's Press 1998 p.157
[xiv] Delorm; p.157
[xv] ibid; p.158
[xvi] Spoto, Donald, *Diana - The Last Year*;; Harmony Books 1997 p. 171
[xvii] Sanction; p.158-9
[xviii] TLC - Mohammed Al-Fayed interview
[xix] Junor, Penny, *Charles - Victim or Villain*; Harper Collins Publishers 1998; p.18
[xx] Sanction; p.167
[xxi] Final Report - Paris Prosecutor's Office; Head of the Prosecution Department at Courts of the First Instance; Examining Magistrates Hervé Stephan and Christine Devidal
[xxii] *TLC* - documentary information
[xxiii] *TLC* - interview with Dr. Martin Skinner.
[xxiv] Anderson; p.191
[xxv] Interview with Mohammed Al Fayed as per his internet site address: www.alfayed.com/indexie4.html, as published to the Internet on October 25, 1998
[xxvi] Spoto; p.172
[xxvii] Sanction; p 251
[xxviii] ibid; p. 6
[xxix] *Newsweek* Magazine; September 8, 1997; p. 33
[xxx] ibid; p. 241
[xxxi] Buchanan, Lyn, *The Seventh Sense*, Paraview Pocket Books, 2003, p. 190
[xxxii] Sanction; p. 17
[xxxiii] ibid; p.17 - 18
[xxxiv] Junor; p. 20
[xxxv] Spoto; p.180
[xxxvi] Junor; p. 22
[xxxvii] *French Final Accident Report* – Conclusionary Statement section
[xxxviii] Lyall, Sarah; New York Times; December 15, 2008

Evidential Details

Part II

What you are about to read is the Remote Viewing Data the Intelligence Community would have received had they tasked this event in the interest of the People of the United States of America.

Evidential Details

Military Targeting Follow-up

"Det G's[1] viewers looked at projects ranging from the status of a cement plant in a hostile country to the location of Soviet troops in Cuba. Important North Korean personalities were targeted, as well as underground facilities in Europe, chemical weapons in Afghanistan, the presence of electronic bugs in the new U.S. embassy in Moscow, the activities of a KGB general officer, a missing U.S. helicopter, tunnels under the Korean Demilitarized Zone, and numerous buildings whose purposes were unknown to U.S. Intelligence."

"But frequently we never learned how close we had come to the truth, how helpful we had been, or even what we had been looking for in the first place. The targets were sometimes so highly classified that substantive evaluations could not be provided."[i]

---

[i] Smith, Paul H., *Reading the Enemy's Mind – Inside Star Gate, America's Psychic Espionage Program*; Tor Non-fiction, 2005; pp. 114-115

---

[1] Det G [Detachment G] was the remote viewing program's code name as it was changed from Operation 'Gondola Wish' to Operation 'Grill Flame'. These were the viewers to make the Army's cut between December 1978 and January 1979. "The Army Chief of Staff for Intelligence, Major General Thompson, officially decreed that the program name, embodied in Det G, would be the focal point for all Army involvement in parapsychology and remote viewing." Op cit. Smith above.

Evidential Details

Greatest Mystery

of the

Civil War

**Leading to America's
Bloodiest Day, the
Emancipation Proclamation, and
European non-recognition of the South**

Solved

"The causes of war are the same as the causes of competition among individuals: acquisitiveness, pugnacity, and pride; the desire for food, land, materials, fuels, mastery. The state has our instincts without our restraints."

Will & Ariel Durant

The Lost Order

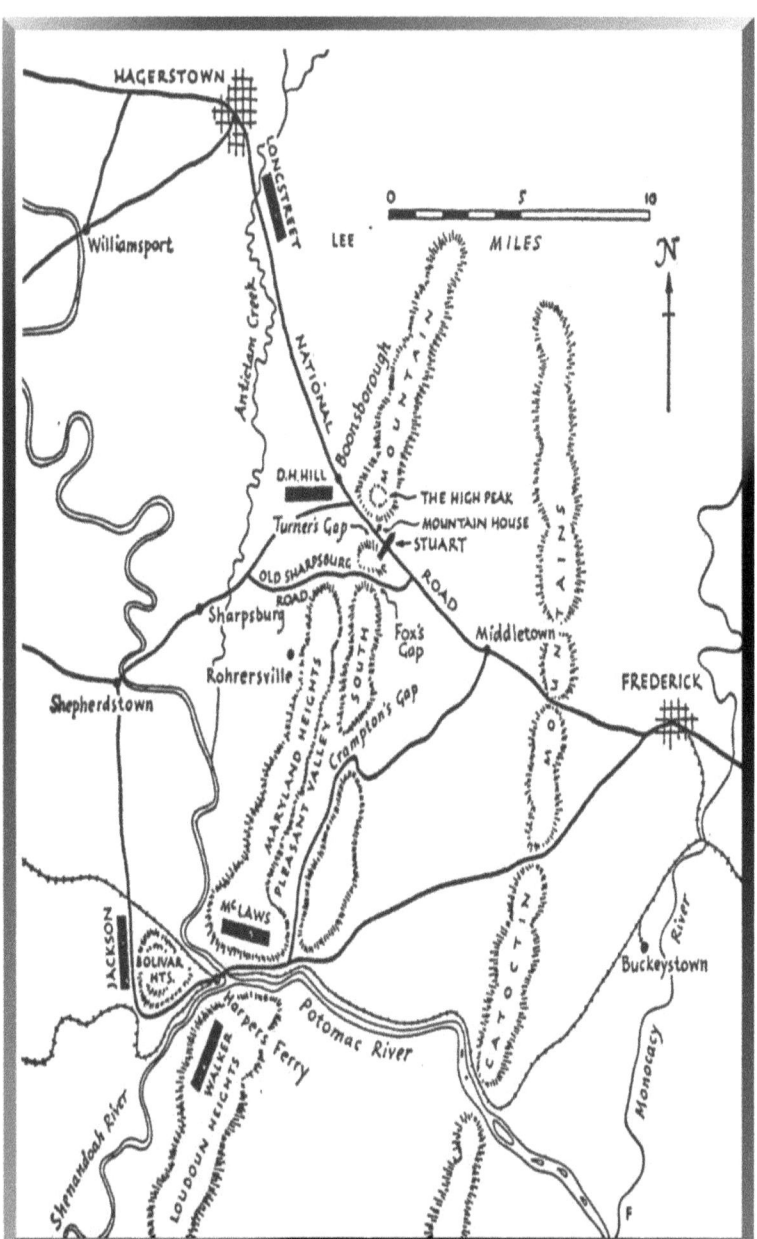

Reference map. Starting at the Monocracy river side campgrounds southeast of Frederick, Maryland (right edge), this map details the mountain ranges, their passes, rivers, and each Southern General's destination as per the Lost Order.

Evidential Details

It has been said success has a thousand fathers while failure is an orphan. But historians have missed a curious exception to this rule as Confederate General Robert E. Lee implemented his bold plan to move the Civil War's carnage north in early September of 1862.

The American Civil War [1861-1865] put brother against brother in a struggle to resolve the nation's States Rights question. From Washington D.C. to Key West, Florida to El Paso, Texas, to Kansas City, Kansas, the Southern States wanted their own nation and fought a long hard war to realize that vision. And it brought innovation.

New non-lethal weapons emerged such as mankind's first tactical use of telegraph, railroads and observation balloons. World War I (1914-1918) style trench works with obstructions were built. Iron clad vessels were launched that made the all world's navies obsolete overnight. With the new scoped and rifled musketry, artillery, revolvers, Gatling guns, field embalming, battlefield photography, and the world's most advanced prosthetics, this war is considered the first modern war. But for the Southerner, it was considered a war for independence.

Summer – State of Virginia. In the midst of a war ending military crisis, Southern General Robert E. Lee assumed command of the Army of Northern Virginia on June 1. By day 25, with a 72% numerically inferior force, he turned back the Federal advances at the gates of the Southern Capital Richmond, Virginia. That week a series of battles, known as the Seven Days campaign, has been compared to the fluidity of World War II battlefields. With his victories over Union Lieutenant General George McClellan, Lee swung his forces north to confront a second Union Army commanded by Major General John Pope [1822-1892]. In his last campaign as an independent Commander, Confederate Major General Thomas J. "Stonewall" Jackson opened the battle on August 9, at Cedar Mountain in Culpepper County.

Now, with a Southern Army that was 87% as large, Lee

The Lost Order

routed Pope's Federal forces at the Battle of Second Manassas. That campaign ended with Jackson slugging it out with Pope in a driving rainstorm at Chantilly Crossroads, on September 1, 1862.

After 90 days in command, Robert E. Lee had moved the battle lines from modern day east suburban Richmond, to the Potomac River northwest of Washington, D.C. Except for Norfolk, the State of Virginia had been virtually cleared of what was considered Federal occupation. Now, amid the Southern press clamoring for quick action to destroy the Union Army, Lee was weighing his options. If the Army moved north into Maryland, he hoped this border state might join the Confederacy. The South should then be able to recruit locally and hamper Washington D.C.'s northbound communications and supply. But after two major campaigns that year, Lee had challenges.

First, there was a need for military re-organization. General Lee added Wade Hampton's Cavalry and some reserve artillery under William Pendleton to his Army. He also added close to three Infantry Divisions in an effort to make up for the loss of approximately 9,000 men in the Second Manassas campaign.

Lee also had to settle command questions. After the recent combat losses, there were not enough Generals to go around. Colonels headed fifty-seven percent of Stonewall Jackson's fourteen Brigades. To ease this situation Generals Dick Garnett, John Bell Hood and A.P. Hill were released from arrest to resume field command.

Absent without leave and straggling was another problem. While considering the move into Maryland, upwards of 30,000 hungry men roaming northern Virginia were directed to meet at Winchester to be united into a cohesive fighting force and to protect against being taken prisoner.

Shoddily manufactured boots, bad food and a quartermaster who could not sufficiently supply the Army aggravated the situation. As Confederate Major General Daniel Harvey Hill wrote: "*The order excusing barefooted men from marching into Maryland*

Evidential Details

had sent thousands to the rear."[i] Soldiers were constantly falling out due to bad nutrition resulting in "GI" symptoms. Compounding the problem was that many young officers, whose average age was now in the early 30's, simply did not have experience keeping irregular columns of soldiers closed.

And finally, the Confederate High Command was plagued with injuries. General James Longstreet was unable to wear a boot due to a painful foot blister. Generals Jackson and Lee himself had both been involved in horse accidents. Jackson was thrown from a new horse he received from a pro-Confederate Marylander when his horse went missing. And General Lee's hands were in splints from a fall while trying to restrain his horse Traveller. He was carried into Maryland in an ambulance.

Summing up the situation, Lee wrote to the Confederate President Jefferson Davis:

"The Army is not properly equipped for an invasion of an enemy's territory. It lacks much of the material of war, is feeble in transportation, the animals being much reduced, and the men are poorly provided with clothes, and in thousands of instances are destitute of shoes."[ii]

To counter this Lee authorized the Quartermaster to buy cornfields, orchards and wooden fences for firewood. In his memoirs, Confederate General John Bell Hood, recalled:

"My troops, at this period, were sorely in need of shoes, clothing and food. We had had issued to us no meat for several days, and little or no bread; the men had been forced to subsist principally on green corn and green apples. Nevertheless, they were in high spirits and defiant, as we contended with the advance guard of McClellan..."[iii]

Rejecting the opportunity to lay siege to Washington, Lee decided to try and draw the Union Army out into the open by moving northwest across the Potomac River. As his Army was being reshaped, Lee was analyzing an ambitious plan to split his forces five ways in the face of an enemy twice his size.

The Lost Order

Alfred R. Waud, September 1862, Library of Congress
Union observers take shots as the Confederate Calvary moves inbound (northerly) across the Potomac River at different locations.

On Thursday September 4, Confederate soldiers started to cross the Potomac River up from Leesburg, Virginia at White's Ford. The passage of 45,000 men required approximately three days. "There was no one to stop them on the Northern shore; only wide-eyed civilians looked at the most ragged, dirty and profane men they could remember. Leighton Parks, a Maryland boy of twelve, saw them as 'a hungry set of wolves,' but he admired the horsemanship of (Calvary General J.E.B.) Stuart's troopers;"[iv]

While history records troop jubilation upon crossing the river, and in some towns, what is not recorded is during the next few days attitudes changed. The men understood they were outnumbered in enemy territory and that their supply lines were extended. Their crossing of a major river made timely supply an open question. And their now compromised ability to maintain surprise was a major component of Southern prospects.

Evidential Details

Author
Modern day White's Ferry facing southeast. The four minute trip across the Potomac River allows historians to interpret Confederate troops moving into Maryland (top left) from Virginia.

Once in Maryland, the Confederate General Staff found their reception muted. "A staff officer recorded: *The inhabitants of Maryland whom we met along the road...did not greet us quite so cordially as we had expected...*"[vi] After the Southern troops had gone into camp, 135 years later Joseph McMoneagle sat in front of a double blind targeting envelope to acquire a time and place coordinate concerning a lost packet in time.

McMoneagle - Somewhere in the Northeast. My very first impression is that of circles inside of circles. Usually, for me anyway, this signifies duplicity or deceit. I also get a sense that this wasn't the only time such deceit was used or attempted during a specified period of time. I have a sense that it was lost near to, or on the edge of, a major city, somewhere Northwest of Washington D.C. perhaps.

This is a Confederate camp. I get a sense they are preparing for a large battle, the clashing of nearly two complete armies. They are bringing so many troops into the area,

---

[1] An Adjutant-General's report from Washington DC., dated November 9, 1880, stated Maryland furnished 46,638 Union troops; Southern enlistment's from the 1862 Maryland Campaign are estimated at 200.

The Lost Order

they are nervous that (their) **troop movements will be known, or that spies might be in the area. A major battle is only a few days away and has something to do with a small town with flanking... positions that run parallel to a road.**

The small town is Sharpsburg, Maryland. The parallel road is the Hagerstown Turnpike.

General Lee's plan was to draw Federal forces out of the heavily fortified Washington D.C. He would then strike General McClellan somewhere in western Maryland with a rested, re-organized and re-equipped army via the Shenandoah Valley. But before he could do this, he needed to reduce the Federal commands at Harpers Ferry and Martinsburg in modern day West Virginia, while the main army lay screened behind the Catoctin and South Mountain ranges.[2]

Lee finalized his plans on Tuesday, September 9. The Army of Northern Virginia, previously of smaller independent commands, was to coordinate across two rivers to surround a division of 12,500 Union infantrymen accompanied by 1,500 cavalry. The target, Harpers Ferry, in what was then the State of Virginia, had already changed hands twice. Lee's strategy, outlined in *Special Order 191*, dispatched 26 of the Army's 40 brigades twenty-five miles apart in the face of the Union Army's four Corps. As a result, there was no time for a siege; Harpers Ferry had to be taken outright. General Thomas J. "Stonewall" Jackson was ordered out for this complicated task.

With a ~45,000 vs. 87,000 troop strength inferiority, Lee's orders were bold. He was counting on the newly restored Union General George McClellan's cautiousness. Lee knew that dividing his forces, in the face of almost 2 to 1 odds is how armies are "defeated in detail." In military circles, this force separation is

---

[2] Virginia's western counties gathered in Wheeling to nullify Richmond's secession act of June 11, 1861. In April 1862, a new constitution was drafted and a separate Governor, Francis H. Pierpont, appointed. On April 20, 1863, President Lincoln proclaimed the State of West Virginia, to be effective in 60 days, making it the 35th State. The capital of Charleston was not announced until 1885.

Evidential Details

known as a violation of Concentration of Mass. However, once decided upon, it is vitally important to maintain the principle of Security. Should this be compromised, the principle of Surprise is violated. Surprise was the critical component of the Confederate plan to make his Maryland campaign a success. Once Surprise is compromised by a lack of Security, Maneuver is reduced. Should a superior enemy learn about a smaller enemy's force divided, throughout history that force, and many times the cause, is lost.

McMoneagle - I believe within two, possibly three days time, they were moving to join a battle... This was a major battle that involved at least four divisions of men, maybe more, spread across a battle line at least ten miles wide and six to ten miles wide on both sides.

The town of Harpers Ferry sits at the confluence of the Shenandoah and Potomac Rivers. Once Lee was north of the Potomac, the elimination of this Federal garrison became essential to maintain Southern lines of communication. After consultations with his Generals in Frederick, Maryland, Lee's *Special Order 191* was issued indicating where each commander was to deploy for the week. Paraphrased, the orders specified:

- General Thomas J. Jackson [1824-1863] was to seize the Baltimore & Ohio Railroad and to drive the Martinsburg garrison (2,500 men) south to Harpers Ferry. Then on to Harpers Ferry, attacking by land from Bolivar Heights in today's West Virginia. Jackson's men would have to march and fight over fifty-one miles in two days to be there on the third.
- General Lafayette McLaws [1821-1897] was to attack Harpers Ferry from Maryland Heights on the Maryland side of the Potomac River. He was to report to Jackson.
- General James Walker [1821-1893] was ordered to attack Harpers Ferry from Loudon Heights across the Shenandoah River from Virginia. He was to report to Jackson.
- General James Longstreet [1821-1904] was to occupy Hagerstown, Maryland and remain in reserve.

The Lost Order

- General Daniel Harvey Hill [1821-1889] was ordered to bring up the rear moving to Boonesboro, Maryland, with the supply trains and cannon.

As the last division to cross, Confederate Major General D.H. Hill's men went into bivouac east southeast of Frederick, Maryland on the Monocracy River. Once drawn up, General Lee sent his orders via standard military courier confident he would retain the advantage of Surprise. As ordered, Southern arms moved the next morning separating across Western Maryland.

McMoneagle – I have a strong feeling of anxiety within the camp. I think these men know they are outnumbered and are nervous about having crossed a river and then camping in a place that essentially leaves it at their back. Very little sound in the camp - - like they are trying not to make any/or as little noise as possible. The weather is cool at night, but somewhat warm during the day - maybe the beginning of the Fall? At least that's what the weather feels like.

Wood smoke in the air. Some food smells. Smells like morning food versus later day food. Heavy mist in the air, maybe ground fog. Damp and cool [near high 60's F]. Sounds of breaking camp. Lots of muffled sounds, as though everyone is trying not to make any noise; Horses being saddled, traces being hooked up on wagons. Can hear orders being given in the distance, as men are marched away.

On the move, one Southern General recalled:

<u>D.H. Hill</u>: "*The Confederates, with more than half of Lee's Army at Harpers Ferry, distant march of two days, and with the remainder divided into two parts, thirteen miles from each other, were in good condition to be beaten in detail, scattered and captured.*"[vi]

McMoneagle - The Confederate soldiers, who were camped near the trees where the packet was found, dismantled their camp there and were out of the area within two to three hours of the man leaving who had actually lost the pack-

Evidential Details

et. It was also lost early in the war.[3] **Either just prior to** (Antietam) **or... following a major battle** (Second Manassas). **Once the camp was packed up and loaded on wagons and mules, they moved the camp in what feels like a Northeast... direction, towards a river** (to the Monocracy River's Crum Ford).

The Southern population had been impressed with Robert E. Lee. "There was already talk of invading the North to teach the Yankee aggressors a lesson. The end of the war seemed near in these first days of September, 1862."[vii] As D.H. Hill's division disappeared west through Frederick, M.D. and into the mountains, the Confederates now held the high ground. Screened by two mountain ranges, and a superb Calvary, the undefeated Robert E. Lee led, man for man, the toughest and most dedicated veteran soldiers ever deployed in the Americas. This Rebel Army had slipped northwest of Washington D.C. And from the mountain tops, they could monitor Federal activity.

* * *

Ever the dramatic, Union Lieutenant General George Brinton McClellan penned a letter to his wife Mary Ellen from his office in Washington D.C. just after 11:00 a.m. on September 5, 1862. "...*again I have been called upon to save the country - the case is desperate... - Truly God is trying me in the fire...*"[viii]

General McClellan had just been directed by President Abraham Lincoln to resume Eastern Theater command the day after the Confederates started their move into Maryland. But the 36-year-old was uncertain how to manage in the face of his recent week long military reversals around Richmond, Virginia.

Though in command of a larger and better equipped fighting force, McClellan worried about what Robert E. Lee would do next. But convinced he was history's "Man of Destiny" called to save the Union, McClellan continued his bold offensive measures

---

[3] The packet was lost in the eighteenth month of a forty-eight month war.

The Lost Order

Mathew Brady or assistant - The National Archives
Union Theater Commander General George McClellan [1826-1885] Commander of the Army of the Potomac and the Democratic Party's 1864 Presidential candidate running against Abraham Lincoln.

Evidential Details

Julian Vannerson, 1863. Library of Congress
Confederate General Robert E. Lee [1807-1870]
Commander of the Army of Northern Virginia.

correspondence with President Lincoln, while he apprehensively paced his office.

A veteran Confederate army was on the move and the populace was uneasy. Alarmist fears of a Southern invasion all the way to Baltimore, Philadelphia and even New York were voiced.

The Lost Order

With a confident enemy moving toward the Pennsylvania border, a stunned Governor Andrew G. Curtin became extremely concerned about Southern intentions.

Aside from not knowing where the enemy was, McClellan also had organizational problems. First, he had to merge his defeated Richmond (Peninsula) Campaign forces with a thoroughly demoralized Union Army of Virginia after their defeat at Second Manassas. Additionally, he had to train, equip and organize no fewer than 35 new regiments of 500 men each. This had to be done in extreme haste and it was a credit to McClellan he organized as quickly as he did.

But McClellan's biggest problem was intelligence gathering. He was up against Western Maryland's topography of Parr's Ridge and the Catoctin mountain ranges in a horse and buggy era. This array of parallel Appalachian hills hampered horse mounted intelligence gathering. These responsibilities had been recently delegated to Cavalry Brigadier General Alfred Pleasonton [1824-1897]. Hard working and a fighter, Pleasonton was nonetheless unable to penetrate the Rebel Cavalry screen.

Having lost the initiative again, General McClellan needed information on three issues: the Rebel army's whereabouts; General Lee's intentions; enemy numbers. As he stewed, he became concerned about the implications of a misstep leading to another military disaster. Having taken Robert E. Lee's repeated beatings with smaller forces, he fretted about:

"*a gigantic rebel army*" worrying that, "*Every other consideration should yield to this; and if we defeat the army now arrayed before us, the rebellion is crushed. But if we should be so unfortunate as to meet with defeat, our country is at their mercy.*"[ix]

Consider his personal drama. With a confident and undefeated Confederate Army screened and on the move somewhere northwest of Washington D.C. the United States, and the Lincoln Administration's, future was on the line. On the other hand,

Evidential Details

as far as the President was concerned, it was time Theater Commander McClellan issue the orders to put this rebellion down and restore the Union.

McMoneagle - McClellan - I would guess that he was one of the top four or five Union Generals. I have a sense that he eventually got into trouble as a result of his 'fear of deceit -- or caution' whichever way you want to look at it. He probably wasn't a very decisive person. Competent overall, but not very aggressive. I also have a distinct feeling or sense that this particular Union commander was particularly cautious anyway. Always looking for deceit, duplicity, or manipulation where there was none to be found. I get a strong sense the Union General was a real screw up in that regard - as we say in the military - a man of no vision. Gutless.

"In the Peninsula (campaign) Lee had out generaled McClellan in large part through his understanding of the personality ...of the Commander."[x] Psychologically, for McClellan, staving off defeat would be victory. So, when one of Lee's top secret orders packets went missing, both armies became engaged in a desperate struggle that would determine the timing of the elimination of the institution of slavery in the United States.

Cautious, General McClelland did order troops forward to the unincorporated community of Ijamsville, Maryland seven miles east southeast of Frederick, M.D. In 1831 one of the town founders, Plummer Ijams II, had granted a right of way to the Baltimore and Ohio Railroad to construct a route through his town as long as they agreed to build a depot. It was on the grounds close to those rails that the Union's XII Corps bivouacked on Friday night the 12th.

* * *

On Saturday morning September 13, Union 12th Corps had orders to advance from Ijamsville northwest along what is today Reich's Ford Road east of Frederick. As the first Union men to move into the area since the Rebel departure, Federal scouts were detailed out to D.H. Hill's old campground before dawn.

The Lost Order

Marching from Rockville, Maryland (lower right corner), the blue line shows the 12th Corps advance to confront Confederate forces northwest of Washington D.C.

McMoneagle - Union scouting - Some men were actually sent to check out the old (Confederate) camp area.

Evidential Details

These scouts or outriders were deliberately sent to the old encampment in order to collect as much intelligence as they could about strength, health and welfare of the enemy. They did this to determine approximately how many men were in the camp, how many horses, to look for garbage burial sites to see how well the men were eating, count wagon wheel ruts - - to estimate cannon, etc; General intelligence collection on the enemy.

It was during this mission that Union scouts found an "envelop" next to a tree. Inside was General Lee's *Special Order 191*. This discovery was to determine the outcome of Lee's Maryland campaign. Where, when, how and why the packet was found have been historical controversies ever since.

As the mystery has been recounted, General Lee's dispatch was found in different places in different ways. For instance, in *Stonewall - A Biography of General Thomas J. Jackson*, author Byron Farwell described this event:

"Private Barton Warren Mitchell of Company F, perhaps <u>looking for some firewood</u>, found the envelope which to his delight, contained three cigars. He at once showed his find to First Sergeant John McKnight Bloss ...and they were astonished to see that it also had Confederate orders... With astonishing rapidity the orders were passed up the chain of command, from company commander Captain Peter Kopp, to regimental commander Colonel Silas Colgrove, to the division headquarters of Brigadier General Alpheus Starkey Williams..."[xi]

Afterward Colonel Colgrove stated he received the packet directly from First Sergeant Bloss, not Captain Kopp. In an unusual wartime development, the Lieutenant, Captain and Major's ranks were bypassed. This was an indication of Bloss's time concern.

McMoneagle - Almost dark. Feels like it is going to rain; overcast and slight wind. Absolutely silent. No talking, just the sounds of a horse coughing, or hoof striking the ground, now and then. No more than twelve men on horses, standing around, while a man is going through the packet to

The Lost Order

see what he's found. Man who found the packet has a civilian shirt on, so there is no rank. I believe they are all dressed as civilians, although I have a sense they are scouts for a Union unit somewhere to the northeast or east-northeast. They found the packet under a tree near where the camp had been.

The lost object itself is cream or buff colored. Approximately 6 - 8 inches long; 3.5 to 4 inches wide; 1 to 1.5 inches thick. It may be made of cloth or skin, my guess being the latter as there are specific kinds of folds in it, which are complicated and deliberate -- something akin to origami [intricate folds]. There is paper inside wrapped around something soft. The object(s) I know now to be cigars, makes perfect sense. I was translating them as hotdogs, which of course didn't exist back then. A real problem for me. The exterior wrapping was a soft skin, probably deer hide which was rolled over and tied shut. I believe the contents were further wrapped in papers that were important to the man who owned the cigars (General Lee).

The lost object was actually about palm size. It was found at the base of a small tree, where it had apparently been dropped or left on the ground. The pouch originally belonged to the Confederacy. When found, it was found by Union soldiers. It was probably Indianans. So, if it was found by soldiers from an Indiana outfit (27th Indiana; Gordon's Brigade; 1st Division of the 12th Corp), then it was found by Union soldiers. Probably scouts or flank riders -- that is men who... scouted ahead or to the side of a division or larger...

Sergeant Bloss's scouts worked north along the Monocracy River to meet up with the infantrymen advancing to Crum's Ford. The memoire of Private Edmund R. Brown, 17, recounted the infantry march along Reich's Ford Road to cross Crum's Ford on their way to Frederick.

"On the 13th we moved by the direct road to Frederick." "There being no bridge over the Monocracy, on this road, we forded the

Evidential Details

Courtesy Global Geo Eye, Commonwealth of Virginia, U.S. Geological Survey, Map Data ©2012 Google

The blue line shows Twelth Corps 4.5 mile morning march. The 27th Indiana was at the point. The white arrows show where Sargent Bloss's scouts rode to inspect the former Confederate campground. The star (left edge) shows officer's tent locations.

The Lost Order

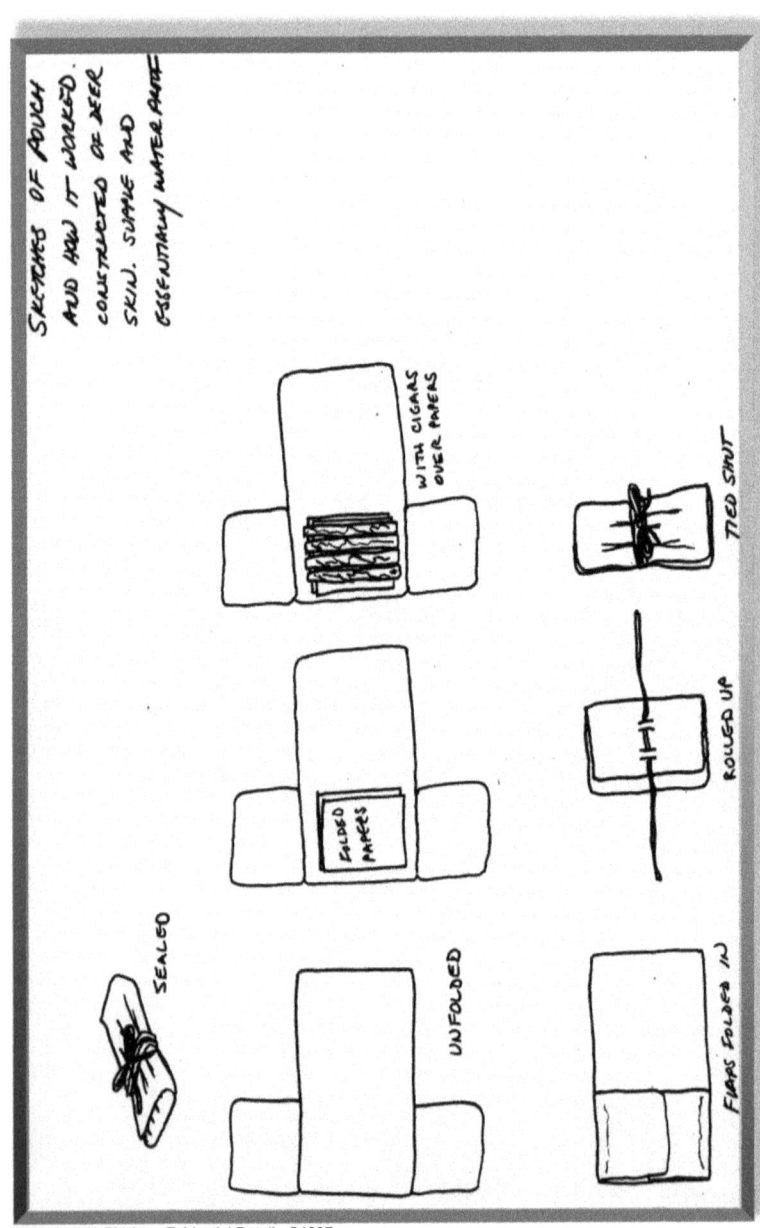

McMoneagle RV Art – Evidential Details ©1997
RV drawing of the Special Orders packet, showing cigar placement. The upper right notes read: *"***Sketches of pouch and how it worked; constructed of deer skin. Supple and essentially water proof.***"*

Evidential Details

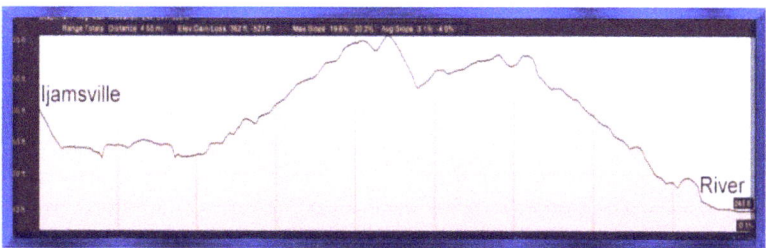

2012 Europa Technologies; 2012 Google; Image U.S. Geological Survey

Marching left to right at less than 4 mph (6.4km), this first terrain analysis demonstrates why it took all morning for the 27th Indiana to cover 4.5 miles. (7.2k) The men had to march against a 3.1% average slope up to a 19.6% grade at the peak of a 523 (159.4m) foot hill. On the west side, they had a 4% average downhill with a 20.2% maximum slope. It slowed the Union advance that morning.

stream." "But with skirmishers still in our front *we moved on* and finally halted in a clover field, adjoining the city on the south."[xii]

In the Pulitzer Prize winning *Landscape Turned Red – The Battle of Antietam*, author Stephen Sears recounted the occasion:

"...The Indiana boys found a comparatively unspoiled section of meadow along side a rail fence and settled down to boil their coffee and relax. Sergeant John M Bloss and Corporal Barton W. Mitchell of company F were chatting idly when Mitchell noticed a bulky envelope in the tall grass nearby and picked it up. Inside was a sheet of paper wrapped around three cigars."[xiii]

In *The Secret War for the Union*, Edwin Fishel wrote:

"McClellan was referring to the greatest intelligence find of the war --- a copy of *Special Orders No. 191* of the Army of Northern Virginia. Wrapped around three cigars ...the order was found by Corporal Barton W. Mitchell,[4] a soldier in his forties, when his regiment, the 27th Indiana Infantry, camped in a clover field at Frederick that morning. Saturday, September 13."[xiv]

In *Stonewall Jackson - Portrait of a Soldier*, author John Bowers speculated:

---

[4] The man that found the orders - Corporal Mitchell - was severely wounded in the leg during the Battle of Antietam a week later and spent the next eight months in a hospital. Like so many, he died from war wounds three years after the war in Bartholomew, Indiana. After some efforts his wife went without a Federal pension.

The Lost Order

"...a Federal soldier had noticed a package lying in the <u>middle of the road</u>. It had dropped from the <u>back pocket of a Confederate Officer</u>. It was Special Order 191 wrapped around a pack of fine cigars."[xv]

In *The Civil War Papers of George B. McClellan*, Editor Stephen W. Sears stated:

"The copy of his operational plan (was) lost by a <u>Confederate courier</u>..."[xvi]

So, did an Officer, a courier or a soldier drop the packet? Was it found at noon while Union soldiers were looking for firewood? Or, were they idly boiling coffee and chatting by a rail fence? Was the packet found in short clover, tall meadow grass, or in the middle of the road?

Brigadier General Alpheus. S. Williams [1810-1878] was Union XII Corp's temporary commander pending Major General Joseph K.F. Mansfield's arrival. He received the packet. But what emerged from these remote viewing sessions was a new time line regarding exactly when Union scouts entered D .H. Hill's former campground and discovered the *Special Order*. We discovered it was retrieved at dawn, but was reported to have been found at about 11:00 a.m. just as Colonel Colgrove arrived. As sometimes happens in a remote viewing, new information results in a reinterpretation of what has been "known" historical fact.

After the rough half day march, the 27th Indiana Volunteers started to arrive along the Monocracy River before noon. Just after the Colonel arrived, First Sergeant John Bloss handed him the mysterious packet. But with this new time line, there is an approximate six hour difference between when the men claimed they found the packet, and McMoneagle's findings. So we decided to see if there were any evidential details that could clear up the discrepancy.

In the Civil War, scouting parties located enemy positions and reported back on camp conditions. This had to be done before

Evidential Details

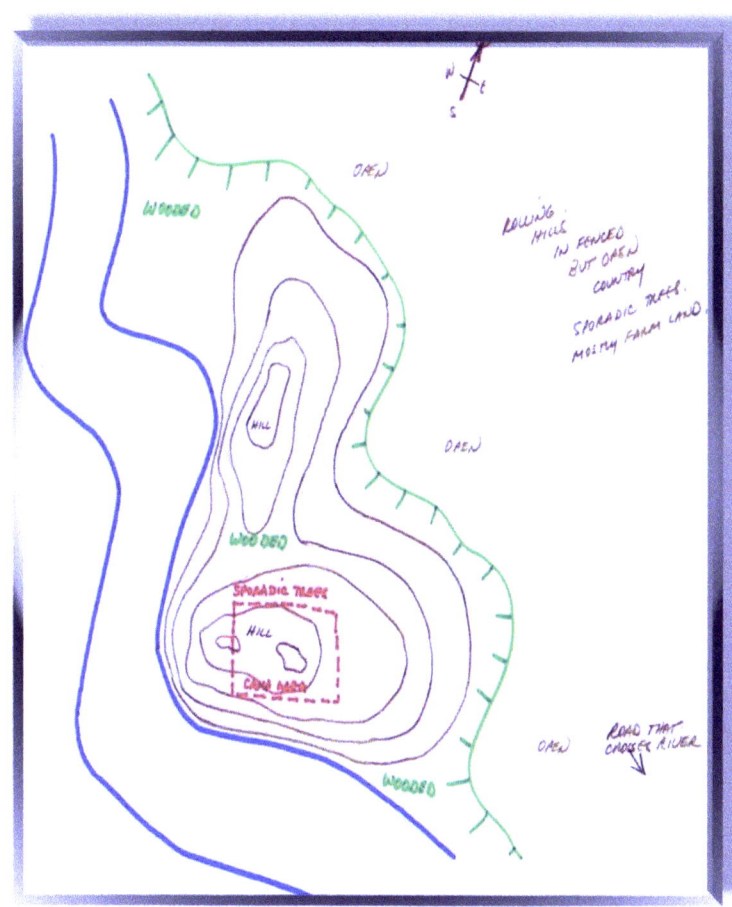

McMoneagle RV Art - Evidential Details ©1997

General D.H. Hill's September, 1862 Monocracy River Bivouac

The Monocracy River southeast of Frederick, MD. Top right the notes read: **Rolling hills in fenced but open country; sporadic trees; mostly farm land.** Notice the sense of stratification including the geographic term "hill" inside the red square.

occupying troop traffic disturbed the grounds. As it turned out, the packet was found at what was the Confederate Officer's hitching tree area on what was a slight rise in the ground. As the Union troops searched "no man's land", it would make sense a scouting party would simply wait for the officer's arrival.

The Lost Order

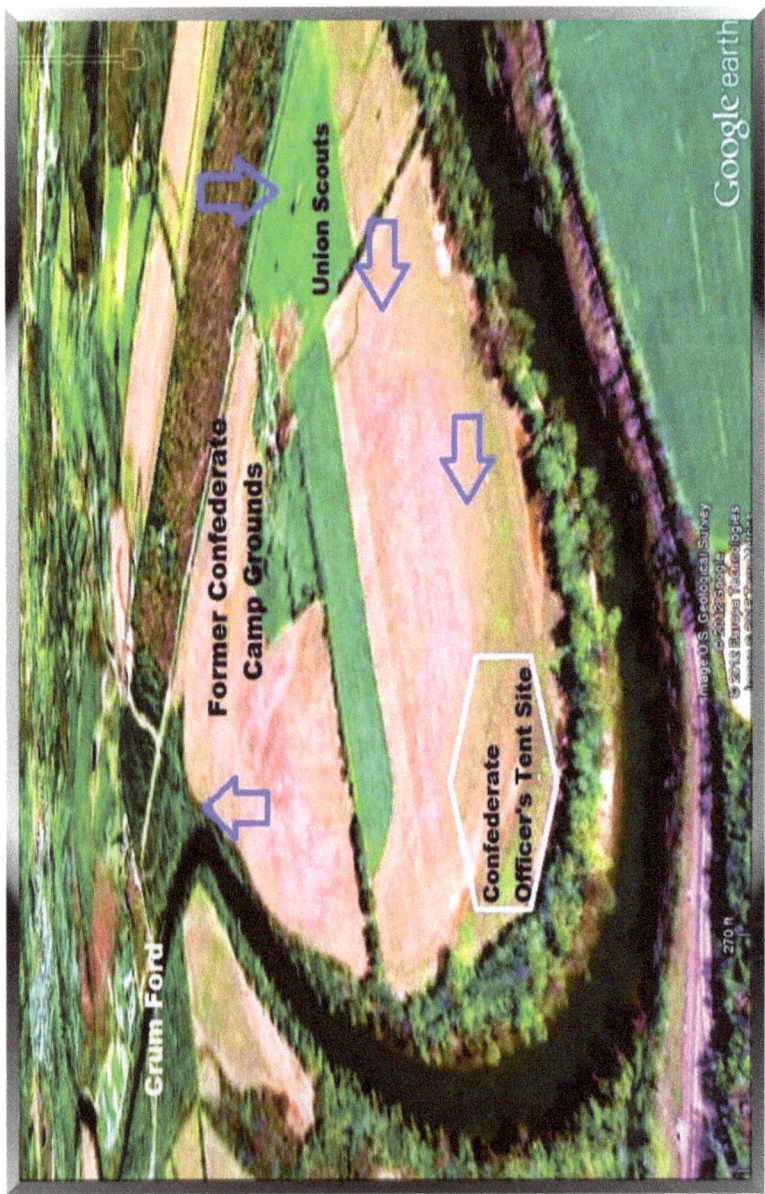

U.S. Geological Survey; Image 2012 Terrametrics; ©2012 Google Earth

From 1185 feet, this graphic confirms the bend in the Monocracy River on McMoneagle's map exists across from Frederick, MD. As the scouts left the road for the vacated camp, this is likely how they approached. The Confederate Officers departed north northeast around the river's bend north to Crum Ford.

Evidential Details

Scout Commander – First Sergeant John M. Bloss

But, in order to believe history, you must accept that the scouting party looked everywhere except the rise with the tree, until six hours of daylight - dawn to 11:00 a.m. - had passed. Then, in an even greater leap of faith, you must believe the men found the packet just as the Colonel arrived.

Sergeant Bloss's scouts had been detailed out to inspect a vacated Confederate camp. The Union Army had marching orders and the First Sergeant had to complete his job, with a squad of men, within a specific time frame. It is not known if Bloss ever received written orders. But as per standard procedure, his men scouted the campground and recorded anything important. Then they moved north and rested as river pickets around Crum's Ford. This accounted for the men relaxing as the advance elements of the 27th Indiana's Company C started to arrive.

McMoneagle was contacted by telephone for clarification on his dawn vs. the midday version. He replied:

McMoneagle - I don't believe that for a minute. The scouts found the packet early in the morning but had other duties before their assignment would have been completed. They scouted the whole area and then waited until the Colonel had moved in and started pitching his tent.

Where this mysterious package was concerned there were no orders. With his limited manpower, Sergeant Bloss chose not to reduce his numbers to deliver a questionable packet. Let an officer make an authenticity decision around noon. He simply held it and waited for the first officer he saw. As he looked the orders over, one cigar was probably smoked.

The Lost Order

U.S. Geological Survey; Image C 2012 Terrametrics; © 2012 Google Earth

The rendezvous area from 760 feet. The orange arrow is Sergeant Bloss's scouting party moving north to rest and wait along the road. The blue arrows are the 27th Indiana arriving (upper right). Where Colonel Colgrove received Sergeant Bloss is unknown. Clearly, the Confederate Officer's former camp grounds were too far away for the orders to have "*just been found*" as the Coronel arrived.

Evidential Details

Interestingly, this historical revision clarifies the officer's future conduct. If the scouts had the packet all morning, they would have had the ability to refer it to Colonel Colgrove upon his arrival. Bloss must have considered it important because he ignored chain of command regulations and walked over to the Colonel as soon as he arrived. We decided to check if the obscure Colonel Colgrove had ever made a statement about this packet.

Remarkably, in a letter dated June 2, 1886, almost a quarter of a century later, it was still clear in Silas Colgrove's mind how quickly he received the package. In a letter, he recalled the day's events:

Colonel Colgrove: *"We stacked arms on the same ground that had been occupied by General D.H. Hill's division..."* Colgrove's next statement remained unacknowledged by Civil War historians at the time of this research.

Colonel Colgrove: "*Within a very few minutes after halting, the order was brought to me by First Sergeant John M. Bloss and Private* (Corporal) *B.W. Mitchell...*"

Almost twenty-four years later Colgrove emphasized the delivery's promptness. The *"within a very few minutes"* time frame suggests Sergeant Bloss conveyed the package upon the Colonel's arrival. Considering the packet important, it took Bloss just a few minutes to walk across the field.

One does not have to be a scholar to question a perfect timing packet discovery dovetailing with the Colonel's riding into the area. But this is what was reported, as the scouts were trapped. To claim to have overlooked the packet all morning, in broad daylight, would have been extreme scouting negligence in the face of the enemy. On the other hand, if genuine, no one would want to be accused of sitting on intelligence like this for half a day. So for the record, Colonel Colgrove thought it best to report they had just found it. Two and one half decades later, the Colonel emphasized there had been no foot dragging where he was concerned.

The Lost Order

The officer receiving the packet from Sergeant Bloss was Union Colonel Silas Colgrove (1816-1907) of the 27th Indiana. Shown here in Brigadier Generals uniform, he was aware of the order's delay and "*immediately*" forwarded the packet to General William's tent.

<u>Colonel Colgrove</u>: "*I immediately took the order to his headquarters, and delivered it to Colonel S.E. Pittman, General Williams adjutant-general.*"[xvii]

McMoneagle – William's Field Tent - I believe the packet was delivered to the Divisional Headquarters for the Union. There were four pickets, or guards, outside the tent at each corner, approximately twenty to twenty-five paces away. There were other officers within 45 to 60 paces of the tent, in a row of tents. It was taken to the General's tent and he opened it inside. There were two other people present inside the tent besides the General, his Aide - and his second in command - a Colonel (Samuel E. Pittman). **The papers in the packet were brought to the General's attention by the officer in charge. He was then dismissed from the tent.**

After reviewing the document, General Williams was not

Evidential Details

sure what to make of it. Here was the information the Federals could not get. Williams was holding Lee's secret Maryland campaign orders as if he was a Confederate General himself. But were they genuine? What happened next ranks as one as one of military history's most astonishing coincidences.

Michigan's Brigadier General Alpheus Starkey Williams (seated center) and staff.

Before the war General Williams, Colonel Pittman, and since June 2, General Lee's Chief of Staff and Acting Adjutant, now a Confederate Colonel, Robert Hall Chilton, had all worked together in Detroit, Michigan. The then U.S. Army Major Chilton had been Michigan's Army Paymaster since July, 1854. Pittman had been a railroad official who did banking business through Chilton. Due to repeated payroll transactions, Pittman was able to identify Chilton's signature as authentic on the spot in the tent. He was likely the only soldier in the Union Army familiar enough with Chilton's autograph to make an immediate identification. His considered opinion had an abrupt psychological impact on General Williams as he weighed the proper response.

The Lost Order

McMoneagle - These papers contained drawings and writing done with a wooden and metal tipped pen. They described a ground layout, which essentially included some kind of a main road, which crossed a main river (Potomac) **and ran through a small town** (Harpers Ferry). **Opposite the town was also a large creek** (Shenandoah River) **that flowed into the river as well. There were specific orders that were written out that accompanied this drawing, showing the placement of troops and equipment, as well as the primary or overall instructions as to what was intended to be accomplished. It is possible that the papers were copied.**

They were. Stonewall Jackson had provided a second copy to Daniel Harvey Hill who, before the Confederate army's reorganization, had reported to Jackson.

But consider the probabilities against instant recognition. In a scenario that no fiction writer would attempt, there was about a one in 159,000 chance that any particular Union soldier in the Eastern Theater that would happen upon the packet that day (87,000 men around Rockville, Maryland; 72,000 men in Washington D.C.). And across the scores of square miles, the "on location" requirement was mathematically complicated by a fluidity variable because the entire XII corps was in motion.

Then there was the officer's rank requirement to be in the General's tent. Next, there was the extremely narrow ten minute time window. Once the orders were forwarded to headquarters, there would be no opportunity to see them again. Pitman's opportunity to identify the signature existed only at the moment Silas Colgrove arrived at General William's tent.

There was also the substitute personnel variable. Chilton's[5] handwriting was spotted because he was substituting for

---

[5] Robert Hall Chilton [1815-1879] has been accused of the orders loss. As Lee's Acting Adjutant, the theory is that, a traitor, he enlisted a courier to drop the packet where it was certain to be found. The allegation's basis is Chilton's 1863 mishandling of Confederate General Jubal Early's discretionary retreat order at the Battle of Chancellorsville which could have caused a disaster for Confederate arms.

Evidential Details

Lee's Adjutant Walter Taylor. And unlike Taylor, Chilton, 47, did not keep up the courier log to confirm if a signed receipt had been returned. Mixed together the probabilities against confirmation, at the point of decision, become nearly impossible to calculate

Because of the ramifications of finding the Special Order, what happened inside General William's tent has always been of unobtainable interest to Civil War buffs - an encounter forever lost. The packet contained a dispatch marked Confidential. Dated September 9, it was entitled *"Headquarters, Army of Northern Virginia, S.O. 191. Official, R.H. Chilton, A.A.G."*. This order was addressed to Confederate Major General Daniel Harvey Hill.

McMoneagle – In William's Tent - The General and the Colonel (Pittman) **were the only ones to discuss the contents, and both were decidedly wary of the information contained within the paper[s]. They both felt that it was a Confederate ruse. The Colonel advised not to share knowledge of the papers with higher headquarters, but the General felt the information should be known. He forwarded it with a personal note that recommended it be taken with mistrust.**

Colonel Pittman's behavior suggests Colgrove told him of the scout's delay. First, he pointed out his prompt packet receipt. Then he mentioned his immediate delivery to Williams. But more importantly, he then recommended his commanding Officer, "…**not to share knowledge of the papers with higher headquarters**..." This appears to be defensive behavior regarding the war's most monumental piece of intelligence. Why should Pittman care what the General did with it? It was because he made the identification.

Consider the alternatives. If Headquarters thought this information genuine, it should be what an Officer would want to take credit for. But on the other hand, if Pittman was aware that the scouts sat on the orders, he had every reason to believe he had been handed the proverbial "cans of worms." If genuine, there was reason to fear an Army Inquiry surrounding the circumstances of how, when, where and by whom the packet had been discovered.

The Lost Order

No one would want to admit his men took no initiative to bring information like this forward for half a day.

The National Archive

A Union General's field tent. One can image two officers standing, looking at the lost orders packet. Here President Lincoln confers with General McClelland after the Battle of Antietam. McClellan was subsequently relieved of command for failure to pursue in force as the Confederates returned to Virginia.

Evidential Details

McMoneagle – In the case of this target, it was not only something that was lost, but it was also found. I believe both the 'losing of something' as well as the 'finding of something' was honestly done. It was accidentally left in a place that was commonly used or common to use by both the Union as well as the Confederacy...

The contents are critically important for some reason. I believe the documents were passed to a divisional commander (Williams) who probably passed them to a ranking General in the Union Army (McClellan). This is probably a large Union force -- Army size.

After getting the orders packet in a <u>very short time</u>, and taking it <u>immediately</u> to Williams' tent, Colgrove emphasized:

<u>Colonel Colgrove</u>: "*It was <u>at once</u> taken to General McClellan's headquarters by Colonel Pittman.*"[xviii]

In his note to Army Headquarters, Williams made sure to mention a corporal found the orders.

<u>General William's Note</u>: "*I enclose a Special Order of General Lee commanding Rebel Forces which was found on the field where my corps is encamped. It is a document of interest & is also thought genuine. The document was found by a corporal of the 27 Ind. Rgt, Col. Colgrove, Gordon's Brigade.*"[xix]

Normally, a General would want to come forward with "look what my outfit found" as a morale builder after a terrible summer of defeats. Maybe the Corporal and the Sergeant should get leave or an extra stripe for meritorious conduct. The whole regiment could take pride in having been the ones to discover General Lee's battle plans. According to the Commander in Chief:

<u>General McClellan</u>: "*Whoever found the order in question and transmitted it to the headquarters showed intelligence and deserved marked reward, for he rendered an infinite service.*"[xx]

But even though his letters home became a book, A.S. Williams never wrote or spoke publically about the lost orders details. Remarkably, XII Corp. wanted the whole matter dropped

The Lost Order

and recognition for a job well done was never mentioned - even after the orders were verified to have been genuine.

McMoneagle - I still keep getting circles within circles, which usually indicates deceit or intelligence mumbo-jumbo... It's as if he is deliberately hiding. So, there is a feeling of duplicity. There is something very 'spooky' or 'undercover' about this object. An object of intelligence or intelligence derived. This isn't the first time, nor will it be the last time, that simple acts are twisted through perhaps disinformation.

It now appears clear the "fatherless" Lost Orders were in Union hands the entire morning. The behavior indicates the Coronel knew the orders had been sat on. One can imagine Pittman's chagrin when he realized his signature identification meant any further effort to prevent the orders from moving up the ladder would seem suspicious. This packet was so important that:

<u>Colonel Colgrove</u>: "...*had it not been for the finding of the lost order, the battle of South Mountain, and probably that of Antietam, would not have been fought.*"[xxi]

And what about General Alpheus Williams?

McMoneagle - I have a sense that he did not do well physically. ...my sense is that he did not suffer from war wounds or anything like that, but was generally subject to quite a bit of sickness in his years during the war as well as immediately following it. I think he died from natural causes within twelve years of the end of the war.

A.S. Williams was not wounded. He died as a Michigan Congressman of a stroke in Washington D.C. thirteen years after the war, on December 21, 1878, at age sixty-eight. A Brigadier General for the whole war, he remains obscure because he fought in so many lost Eastern Theater battles. But the reason for his silence about the orders packet has finally been cleared up.

While the Confederacy launched no internal investigation, *Special Order 191's* loss forced Southern Arms to fight out of alignment from behind Antietam Creek on September 17. It remains the

Evidential Details

bloodiest day in American History, and directly affected the timing of the Emancipation Proclamation. These then are the amazing circumstances leading to America's elimination of the institution of slavery. But was not necessary to have remote viewed this target for historians to have realized that the Lost Order became history's foremost example of a spectacularly successful orphan.

* * *

Incredible opportunities mixed with menacing consequences. At approximately 11:30 a.m., Adjutant Seth Williams (1822-1866) handed General McClellan the packet. Upon looking at the orders, McClellan exclaimed out loud in public, "*Now I know what to do.*"[xxii] He then broke off a meeting he was having with some local people about Federal troops in Frederick, Maryland and retired into his field tent. But one of those people was a Southern spy who went directly to Confederate Cavalry General JEB Stuart that evening, and relayed the story of McClellan's abrupt reaction to the Adjutant's packet transfer.

General McClellan was now in procession of his enemy's orders as if he was one of Robert E. Lee's Generals. He now knew Lee's army had been separated for over 72 hours; that Martinsburg and Harpers Ferry were targets; of the alternative regrouping locations; and each enemy General's marching route.

But even with this, as was his character, McClellan developed reservations. He hesitated until 3 p.m. when he asked his cavalry commander to see if the lost dispatch "*order of march*" had been followed. Finally, at 6:20 p.m., he signed orders for Generals Franklin and Couch to move forward in the morning. Thus the precious hours of noon, to after 6 a.m. the next morning, had been squandered by the North - but not by the South.

In retrospect, few commanders in history have ever been given an equivalent opportunity to destroy an enemy as had been given George McClellan. With a quick and massive response, anyone, with 2-to-1 odds, would be able to defeat an opponent

The Lost Order

with his forces split five ways. The rebellion, east of the Appalachian Mountains, should now be over by the end of September. "The lost order represented the opportunity of a lifetime."[xxiii] Robert E. Lee, and his Army of Northern Virginia, would now face the consequences.

At this point speculation is rife about delays. If marching orders had been issued to move to the mountain passes at 12:20, rather than 6:20 p.m., the next day's attack could have commenced at dawn, rather than the next afternoon. But McClellan also waited to order more force forward until the next morning.

McMoneagle - General McClellan - Not much of a commander actually. I think he erred on the side of being too cautious. My sense is that this general was in fact too cautious all the way around. He I view as a weak and otherwise useless individual. He was, what I would call, a Desk General, preferring to sit behind a desk in Washington, rather than lead men in the field. He didn't have the 'close with the enemy & kill the enemy' fervor that is required in a combat leader. I believe he will take the blame for many of the mistakes about to take place during the battles over the next three days.

Though delayed, once it got under way the Union advance did take Lee off guard. And his timetable had fallen behind schedule. "Not one of Lee's Harpers Ferry columns had conquered 191's stringent schedule. Not one (General) had attained its objective by Friday, September 12. Yet all were approaching their designated targets. Did one day really matter?"[xxiv] McClellan's 18 hour overnight delay proved to be a critical factor in what would become a razor's edge struggle.

The Lost Orders made McClellan one of the most informed commanders in world history, but he could not see it. In *Landscape Turned Red,* author Stephen Sears wrote about how McClellan, "...managed to throw away the glittering opportunities for total victory offered him by the Lost Order..."[xxv]

Evidential Details

McMoneagle - My sense is that while deceit was never intentional by the loss of the object, there was a presumption of deceit that was felt by the Union commander, or decision maker; Sort of a sense that the Confederacy might be setting the Union up. In other words, that there might be a plan initiated to fool someone within the Union camp.

But, while the orders and map were absolutely accurate, they were not believed to be. My sense is that the Union Commander thought it was probably an act of disinformation. There was cause for some to suggest that they might be faked or a ruse to try and get the Union command to make the wrong decisions. So, they were never fully believed.

In *George B. McClellan - The Young Napoleon,* Stephen Sears dedicates a whole chapter to the order. "September 13 was also the day the Federals' dismal intelligence gathering record took a startling upward turn with the discovery of the famous Lost Order." Sears continues, "The Lost Order represented the opportunity of a lifetime."[xxvi] Much has been made of McClellan's "extreme caution" after receiving Lee's orders. This was based on his certainty the Confederate Army was almost three times larger than it actually was. But doubts about his state of mind have always run deeper.

As concerns McClellan's mentality, "Repeatedly he saw divine purpose in his being called to Washington to save the country. At the same time, he continually saw himself a victim of harassment. The personality revealed in all this is a mixture of complexes --- persecution, messianic, Napoleonic."[xxvii] As a result, even with the Confederate battle plans, "...McClellan, by force of his temperament, succeeded in transforming himself into a badly misinformed commander."[xxviii] "...our impression that McClellan was eccentric, quite able to believe the unbelievable, is confirmed. We can give up trying to understand him."[xxix]

McMoneagle – Could this man also have been manic depressive?

The Lost Order

Joseph McMoneagle is the first and only remote viewer to be inducted into the American Psychological Association and has lectured the Psychology Faculty at Harvard. Here his decorated remote viewing capability provides history's first genuinely plausible psychiatric evaluation – on a dead man.

* * *

Facing an unforeseen Union advance, a puzzled Robert E. Lee was forced to issue revised orders. He absolutely had to hold the line at the passes at South Mountain, while Jackson continued on to Harpers Ferry. He sent orders to General Daniel Harvey Hill.

D.H. Hill – *"I received a note about midnight of the 13th from General Lee saying that he was not satisfied with the condition of things on the turnpike or National road, and directed me to go in person to Turner's Gap the next morning and assist (Lieutenant General J.E.B.) Stuart in its defense."*[xxx]

As he approached Hagerstown, Maryland Lee also ordered an increasingly disgruntled General Longstreet to countermarch his two divisions 13 miles to support Hill's men at South Mountain. Riding over to Turner's Gap early on the 14th, Harvey Hill found a serious situation. From his position on the mountain, he looked out at the eastern horizon and was overcome.

D.H. Hill – *"...I had seen from the lookout station near the Mountain House the vast army of McClellan spread out before me. The marching columns extended back as far as the eye could see in the distance; It was a grand and glorious spectacle, and it was impossible to look at it without admiration. I had never seen so tremendous an army before, and did not see one like it afterward. ...here four corps were in full view, one of which was on the mountain and almost within rifle range."*

As the battle commenced, it was a sight for Union soldiers as well. As he approached up the National Road to the pass in the Catoctin Mountains and looked across the valley at Confederate positions on South Mountain, a Union private recalled:

Evidential Details

<u>Miles Clayton Huyette</u>: *"When we reached the summit where we could look down and across Middletown Valley - through the shimmering heat waves we could see lines of infantry - which looked like ribbons of blue - being rushed to the front; the rays of the sun glinted from musket... bayonet and polished brass cannon; batteries in position - in the fields in the valley - firing over the heads of the advancing line of battle; smoke of shells bursting in air at Crampton and Turner's Gaps, and could see lines of smoke in the woods near the summit of South Mountain - where the infantry boys were fighting in the brush among the rocks. It was a panoramic moving picture."*[xxxi]

Hill knew his small force would have to fight alone until he was reinforced. He also knew that what was at stake was the imminent collapse of the Confederate eastern theater war effort.

<u>D.H. Hill</u> – *"The sight inspired more satisfaction than discomfort; for though I knew my little force could be brushed aside as readily as the strong man can brush to one side the wasp or the hornet ...I hoped to be able to delay him* (McClellan) *until General Longstreet could come up and our* (supply) *trains could be extricated from their perilous position."*[xxxii]

"My division was very small and was embarrassed with the wagon trains and artillery of the whole army, save such as (Stonewall) *Jackson had taken with him." "Had he* (McClellan) *made the movement which* (Calvary Commander) *Stuart and myself thought he was making, it was hardly possible for the little force under Lee ...to have escaped, encumbered as it was with wagon trains and reserve artillery. Forming his infantry into a solid column of attack, Lee might have cut a way through the fivefold force of his antagonist, but all the trains must have been lost, -- an irreparable loss to the South."*[xxxiii]

If the Union army had forced its way through any of the gaps, Lee's army could be cut in two, defeated in detail, and his supply trains and cannon captured. But when D.H. Hill got to Turner's Gap Calvary Commander JEB Stuart was gone. Only a half strength 200 man regiment under Thomas Rosser, with an

The Lost Order

artillery battery, remained. He also discovered Brigadier General Alfred Colquitt's men had moved down the mountain and could not participate.

General D.H. Hill's South Mountain battle Headquarters photographed in 2000.

D.H. Hill – *"In fact, after the removal of Colquitt's brigade, about 7 A.M., there was not a Southern soldier at the foot of the mountain until 3:30 P.M., when Captain* (Robert E.) *Park of the 12th Alabama Regiment was sent there with forty men."*

Facing up to a bad situation, Hill opened with cannon when the Union forces came within range. Encountering resistance, Federal troops swung a mile south to Foxes Gap in an attempt to flank the Confederates. Here circumstances became extreme. (See page 59)

D.H. Hill – *"Fifty skirmishers of the 5th North Carolina soon encountered the 23rd Ohio, deployed as skirmishers under Lieutenant-Colonel R.B. Hayes, afterward President of the United States. ...the action began at 9 am between* (Union Major General Jacob) *Cox's division and* (Brigadier General Samuel) *Garland's brigade."*[xxxiv]

Evidential Details

General Cox had a 2 to 1 advantage over General Garland. Worse for Southern arms, Garland was mortally wounded in the midst of the battle at the mountain top. With their command structure in disarray, the Confederate units were mauled by the Federal onslaught.

The Union Army (right edge) fought uphill against smaller Confederate forces.

D.H. Hill – *"Garland's brigade, demoralized by his death and by the furious assault on its center, now broke in confusion and retreated behind the mountain, leaving some two hundred prisoners... The brigade was too roughly handled to be of any further use that day. There was nothing to oppose him (Cox). My other three brigades had not come up; I do not remember ever to have experienced a feeling of greater loneliness."*[xxxv]

Fox's Gap was now wide open for the Federal's to secure the high ground, come streaming through and open the way for the whole Union army. From this point, they could cut off a southbound retreat for a third of the Confederate Army. The game appeared over. But then, in an incredible turn of events:

D.H. Hill – *"General Cox (Kanawha Division) seems not to have suspected that the defeat of Garland had cleared his front of every foe." "It was more than half an hour after the utter rout and dispersion of Garland's brigade when (Brigadier General) G. B.*

The Lost Order

Anderson arrived at the head of his small but fine body of men." "Under the illusion that there was a large Confederate force on the mountain, the Federals withdrew to their first position in the morning to await the arrival of the other three divisions of (Union Major General Jessie) *Reno's corps."*[xxxvi]

"It was incomprehensible to us of the losing side that the (Union) *men who charged us so boldly and repulsed our attacks so successfully should let slip the fruits of victory and fall back as though defeated." "Thus it was that a thin line of men extending for miles along the crest of the mountain could afford protection for so many hours to Lee's trains and artillery and could delay the Federal advance until* (Lieutenant General) *Longstreet's command did come up, and, joining with mine, saved the two wings of our forces. McClellan could have captured Lee's trains and artillery and interposed between Jackson and Longstreet before noon on that 14th of September."*[xxxvii]

In an incredible turn of fortune, Union forces did not move again until 4:00 p.m. But by 3:30 Longstreet's countermarching troops began relieving the Confederates at Turner's Gap. Meanwhile, over at Harpers Ferry, Stonewall Jackson's men began to set up their attack. His men worked around the clock, methodically positioning cannon to reduce the Federal stronghold.

So far, Southern arms had been lucky. But once the Union Ninth Corp had all its men up an all out attack was launched at Turner's Gap. A general battle up the middle, and on both sides, raged until sundown. Though the Confederates held their ground, they were severely outnumbered and could not continue. Lee then ordered Hill to withdraw by the morning of the 15th. But another critical day had been purchased.

Six miles south, lay the most vital pass at Crampton's Gap. A victory here would put Federal forces just five miles from Confederate General McLaws back at Harpers Ferry. Moving into position were the 19,500 men of the Union VI Corps under Major General William Buel Franklin [1823-1903]. His orders were to cut-

Evidential Details

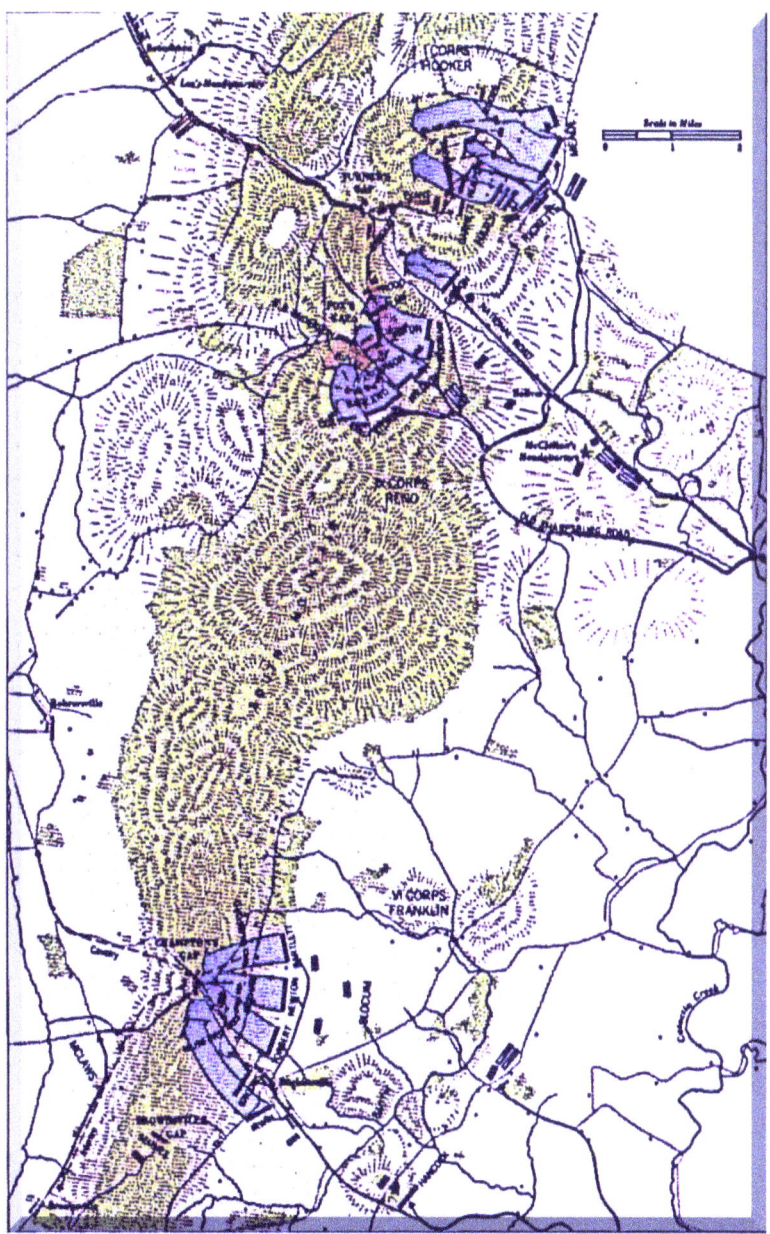

The Union onslaught. *Special Order 191* resulted in battles at **Turner's Gap** (top), **Foxes Gap** (just below) and **Crampton's Gap** (bottom). These holding actions bought time for the divided Confederate Army to re-group behind Antietam Creek.

The Lost Order

off, *"destroy or capture McLaws command and relieve* (Union) *Colonel Miles."*[xxxviii] There were only 1000 Southerners in this fight.

At noon, the forward division under Major General Henry Slocum encountered two regiments under Confederate Colonel William Parnham along with Thomas Munford's dismounted Calvary. Slocum's commanding officer described what happened.

<u>General Franklin</u> – *"The enemy was driven in the utmost confusion from a position of strength and allowed no opportunity for even an attempt to rally, until the pass was cleared and in the possession of our troops." "The victory was complete, and its achievement followed so rapidly upon the first attack that the enemy reserves, although pushed forward on the double-quick, arrived but in time to participate in the flight, and add confusion to the rout."*[xxxix]

At 19 to 1 odds, the Rebels were devastated. And, at Harpers Ferry, Stonewall Jackson had still not opened his attack. How long that battle would last no one knew. The only certainty was that Crampton's Gap had been breached and General McLaws division at Harpers Ferry appeared subject to an overwhelming attack from behind.

Robert E. Lee sensed it was over. The mountain gaps had been breached and the Harpers Ferry assault was two days behind schedule. General Hill had withdrawn from South Mountain. Lee's army was split and demoralized at critical points. Prognosis for the next day - military disaster. So at 8 p.m. on the 14th, revised orders were issued to General McLaws.

<u>R.E. Lee</u> – *"General: The day has gone against us and this army will go by Sharpsburg* (Maryland) *and cross the* (Potomac) *river. It is necessary for you to abandon your position tonight."*

But four hours earlier Stonewall's men had phased in an artillery barrage against the Union garrison at Harpers Ferry. While time consuming, southern artillery had been well positioned in the hills. Then the rebel signal corp. coordinated point blank cannon fire from the surrounding hilltops that raged until sundown.

Evidential Details

The Potomac River in 1862 and the "common center" at Harpers Ferry. The Shenandoah River enters just past the house (middle right edge). The Federal's in town were pounded by Generals Walker and McLaws left and right from the heights across the river. Here the historian can visualize the close to point blank range enjoyed by Southern cannoneers. This camera angle does not show the Federal defense perimeter against Stonewall Jackson's Bolivar Heights land assault which was behind the cameraman.

The Lost Order

"The great circle of artillery (fired) to a common center, while the clouds of smoke, rolled up from the tops of the various mountains, and the thunder of the guns reverberating among them, gave the idea of so many volcanoes."[xl] The Union position was now virtually indefensible. One soldier wrote, "*The infernal screech owls (Confederate shells) came hissing and singing, then bursting, plowing great holes in the earth, filling our eyes with dust, and tearing many giant trees to atoms.*"[xli] Union counter fire was ineffective in its efforts to shoot uphill.

During the night, Jackson moved his cannon forward while, at 8 p.m., Union Colonel Benjamin Davis led his 1500 man Calvary across a Potomac River pontoon bridge and rode 50 miles in 12 hours to escape Confederate forces in the area.

Again at dawn, Southern cannon opened with a vengeance and by approximately 7:30 a.m., Federal artillery shells ran out. The surrender ceremony was completed by eight o'clock. General Jackson left General Ambrose Powell Hill's [1825-1865] division behind to send south the 12-13,000 captured small arms, 200 hundred wagons, 1200 mules and 73 cannon, and hurried to support Lee at Sharpsburg, Maryland. As the Army of Northern Virginia congealed behind Antietam Creek, D.H. Hill, remarked:

<u>D.H. Hill</u> - "*But from whatever standpoint it may be looked at, the Battle of South Mountain must be of interest to the military reader as showing the effect of a hallucination in enabling 9000 men to hold 30,000 at bay for so many hours...*"[xlii]

After the war, Lee claimed he was out to crush McClellan during the Maryland Campaign. Regarding the stakes, General Walker said Lee told him directly:

<u>R.E. Lee</u>: "'*I shall concentrate the army at Hagerstown* (Maryland) *...and march to this point,*' placing his finger at Harrisburg, Pennsylvania. '*That is the objective point of the campaign.*'"[xliii]

The target was the railroad bridge over the Susquehanna River which, if destroyed, would shut down the North's ability to

Evidential Details

move material by rail in an east - west direction. Lee would attempt the same objective again in July, 1863. His efforts culminated in a decisive three day affair in a sleepy little Pennsylvania hamlet called Gettysburg. But in 1862, "...the final unforeseen factor affecting Lee's strategy was the finding of the famous document known to history as the 'Lost Dispatch'."[xliv]

With Northern newspaper stories and testimony before a Conduct of the War Joint Congressional Committee in March 1863, Lost Order speculation abounded. Both sides wanted answers to the intriguing question about how the packet had been lost, and who lost it. So, as this target was submitted to McMoneagle's office in 1997, Civil War historians agreed that, "In the absence of new evidence, it seems unlikely that there will ever be a solution to the mystery of how Order 191 was lost."[xlv] [6]

* * *

Who, how, why, exactly when and even where the orders were lost, has been the Civil War's most significant mystery ever since. Some Historians have even written about the possibility of ending the war thirty-one months early. "Lee called the enemy's possession of the order, *'a great calamity'*"[xlvi]

Some implicate Daniel Harvey Hill simply because he was the addressee. But could we confirm that the man, who for the rest of his life hotly denied losing the orders, was really to blame? The General assures us: "*It is the unusual that impresses.*"[xlvii]

A glimpse of D.H. Hill's life shows that in many ways he was his own worst enemy. He graduated from West Point 28th in a class of 56 in 1842. But he was an instigator. For example, he thought it was a good idea to bypass the Army's chain of command by going directly to Confederate President Jefferson Davis with a

---

[6] The *New York Herald* and the *Washington Star* newspapers published partially correct stories about the lost order on September 15, 1862. The origin of the Congressional leaks has never been determined, but a press connected George A. Custer had access to this information through a Cavalry staff meeting.

The Lost Order

lack of vigilance report on the part of the Cavalry. He then openly questioned the Cavalry's manhood with his standing "joke" of a $50.00 cash reward for anyone finding a dead solider with spurs on. Later, due to Hill's ongoing criticisms, he had to be prevented from fighting a duel with non-West Point Brigadier General Robert Toombs over troop handling issues. In writings to his wife, Isabella Morrison Hill, he referred to Confederate Congressmen as "fools".

The problem was that Harvey Hill knew right from wrong. The youngest of 13 children, he had an intelligently "grim demeanor" with an educatedly derisive wit that he used to expose shortcomings in others. In a military school instructional manner, he liked to point out how a stranger could better himself.

Hill also had impeccable religious credentials as one who could forthrightly instruct another how best to turn their life away from sin. And he was clear on his duty to proffer freelance spiritual guidance to help men save their souls. General Longstreet said, "*He was a bitter, sarcastic critic of the frailties of humans.*"[xlviii]

Elsewhere it was said Hill could even go so far as to be "*eccentric on the verge of wrongheadedness.*"[xlix] In a paraphrased statement from the ever reserved Robert E. Lee it was recounted that he thought, "*D.H. Hill had such a queer temperament he could never tell what to expect from him.*"[l]

It was also because of being a West Point graduate that being under suspicion of losing the order, and then being suspected of lying about it, led to a hapless military career. Because Hill was a fighter, he had hung on at the South Mountain battle. Then he went on with General Gordon to fight in the horrific Confederate center at Antietam Creek known as the sunken road, or "Bloody Lane." But the way he expressed himself was not what military men wanted to hear.

D.H. Hill: "*The battle of South Mountain was one of extraordinary illusions and delusions.*"[li]

It is unclear when Harvey Hill first realized McClellan had been in possession of the lost order. In correspondence Hill wrote,

Evidential Details

"*That an order from Lee directed to me was lost, I do not now doubt.*"[lii] But he would continue to deny having received it. He even had his adjutant Major Rachford sign an affidavit stating he had not received it. Daniel Harvey Hill vociferously protested his innocence. It was not in his character to be wrong or lose face. But still dogged by inquiries over 15 years later, in 1880 he wrote:

D.H. Hill: "*This order...was addressed to me, but I proved twenty years ago that it could not have been lost through my neglect or carelessness.*" Hill also floated the idea of treachery.

What the proof was that you had not done something was unclear. But going forward post war credibility would be essential as he interviewed for the University of Arkansas Presidency [1877-1884] and then, in 1885, President of today's Georgia Military College in Milledgeville, GA.

Near life's end, he was afflicted with painful stomach cancer. General Daniel Harvey Hill, 68, died in September of 1889 - twenty-seven years to the month after the lost orders were lost. On his deathbed his last words were, "*Almost there.*"

* * *

Owing to Hill's staunch denials about losing the orders, and a lack of proof, there have been many theories as to who lost the dispatch. Through the Twentieth century, many have looked elsewhere and substantial research has been conducted on every other connected individual. One plausible explanation has been that Jackson's courier, Henry Kyd Douglas, dropped the orders. So now it was time for McMoneagle to solve this mystery.

McMoneagle - Dress profile - I see a man dressed in Confederate Gray with gold braid sewn into the sleeves and edges... He is wearing a single large gold star on his collar. Man who loses [the] packet has three stars on his collar, single line of braid across both cuffs.

Regardless of rank, a Confederate General's collar eliminates couriers, staff and support personnel. What is confusing is

The Lost Order

that Confederate General's regalia contained one large star in the middle of three, but this did not mean the soldier was a Lieutenant General. Harvey Hill had been a Major General since March 27, 1862. So this single component is inconclusive.

McMoneagle - Shorter than most; about five feet, six inches in height. Sandy colored hair about 145 pounds, green piercing eyes... Whether the man had pale blue or pale green eyes, my color maybe a little off, but that man had pale eyes. He has a heavy ...beard but no mustache. He is riding a "roan" [dark red-brown horse which almost looks black]. He limps.

In *Joseph E. Johnston - A Civil War Biography* author Craig L. Symonds indicates, "Hill was small and fine-boned, with lank blond hair and pale blue eyes."[liii] Available photographs also showed Hill's eyes were so pale they can be observed with archaic 1860's camera equipment. This cannot be said about the other Confederate Generals in the Maryland campaign.

But was there any information about a Southern General with a light build, the lack of a mustache and limping? These were historically unheard of evidential details. While no photographic evidence exists, research did find references to these descriptions for one Officer.

About D.H. Hill, a C.D. Fishburne remarked: "*He was then comparatively a young man, wore full whiskers but no mustache, was slightly built, of serious aspect...*"[liv] So, Hill was known to have differentiated himself from the normal bearded look by shaving his mustache. None of the other Maryland campaign Generals were ever known to have made this shaving effort.

Regarding intermittent stiffness forcing a limp research turned up: "*A slender, well-proportioned man not long past 40, of medium height and erect military posture, except when pain from a chronic spinal ailment forced him to bend slightly...*"[lv]

Bending slightly as he walked would have accounted for the appearance of limping. None of the other Confederate Generals limped. So, what about his personality? The Wilmington Mes-

Evidential Details

senger newspaper wrote about Hill: *"He was what he seemed. There was no hypocrisy or guile or sham about him."*[lvi]

McMoneagle – He is stern, doesn't drink, his only vice being probably smoking a pipe or cigars. He has a reputation of being very "uncompromising" and historically doesn't listen to his men or seek advice. My sense is that he is a deeply religious man... A no nonsense kind of person who is both liked and disliked for his black and white views. No gray areas within this person's thinking; probably aggressive and unforgiving in a sense. The guy who was in charge of the (Confederate) **camp is the man who rides the Roan horse.**

Of the five Generals reporting to Lee that fall, this psychological profile is best fitted to Daniel Harvey Hill and Stonewall Jackson. But Hill presents the best combination of physical characteristics. And so, for the first time we learn how the Special order was lost.

McMoneagle - The man who lost the packet was mounting his horse to join his men who were encamped nearby. Everyone was packing up camp and moving. They were just given orders on where they were moving to and why. The packet was dropped when he was alone but he was in the process of joining up with others. I get a sense that he went to his horse, which was back away from where the tents were being taken down. His was the last horse there.

This object or package probably belongs to a very important individual within the Confederacy; someone who has himself known that he has lost this. He honestly lost the object in question, which was probably some kind of a dispatch or letter. When he mounted his horse, the packet was lost. Packet was lost as the man carrying it was climbing into his saddle. I see him putting it into an inside coat pocket, but missing the pocket all together, having it quickly slide out from inside his jacket while either mounting or riding his horse. He didn't notice the loss of the packet. Once he was in

The Lost Order

the saddle, he joined others who were marching away from or out of the camp area. The packet laid on the ground more than 24 hours...

Hill was viewed limping, signaling discomfort from his spinal ailment when he was handling the packet. Nagging pain made it easier to fail to notice the orders were dropped. It would also make it easier to miss his jacket's inside pocket.

This can be understood by observing a man mounting a horse with back pain severe enough to make him limp. Imagine, with his hand holding the saddle post, Hill was in motion to swing his right leg, shoulder blade, and arm over the horse while attempting to insert a packet into his left coat pocket. If the orders packet did not seat into the pocket properly, it would have slipped down inside as the coat was closed. Most business men, even without a spinal ailment, will undershoot their inside suit coat pocket sooner or later. This is particularly true if they attempt to insert a packet without holding the jacket with their other hand.

McMoneagle – I believe he later thought that he had left it in a tent back at the camp he had left earlier in the day. He regretted losing the packet, but it was not a big deal for him, as he believed it to be safe with someone else at the camp. He knew his orders for the day and the following week, so he would not have needed the packet to get his job done.

The information in the packet was a copy of the same information given to at least a handful of men. Each packet of information contained the specifics for an overall plan of action that would be associated with the coming battle. Ultimately, the packet was not essential to the man who lost it. I have a sense that he was one of the three or four other officers who were given like copies or near similar copies of the same thing.

It was left lying on the ground under a tree approximately twenty-five yards from where the General's and other officer's tents were pitched. The tents were struck and the

Evidential Details

camp moved within two hours of the loss of the packet.

Confederate Major General Daniel Harvey Hill
Shown with a moustache, the historian can confirm his eyes
light shading and how the three stars on his collar were situated.

Civil War Library and Museum

The Lost Order

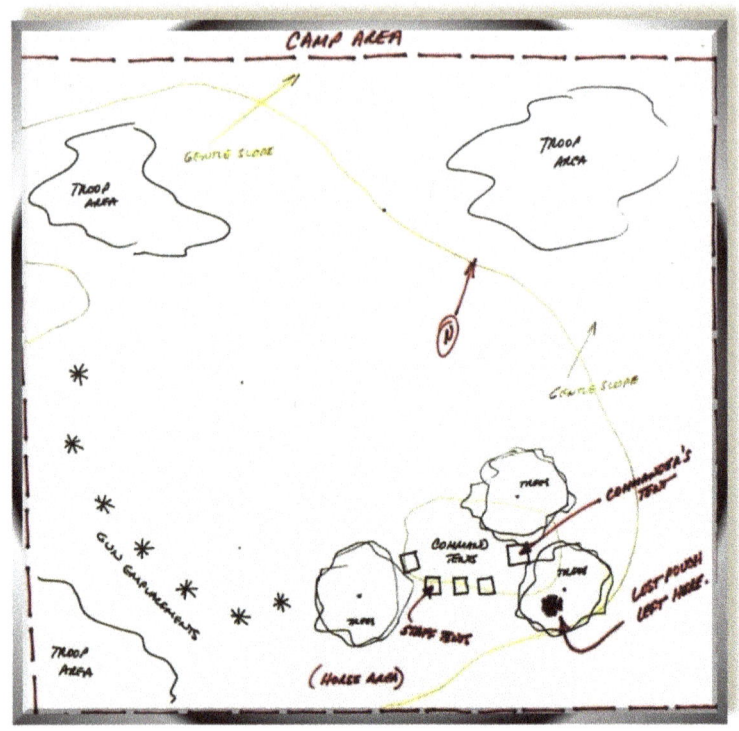

McMoneagle RV Art - Evidential Details ©1997

General D.H. Hill's 1862 Campground Close-up

This remote viewing sketch shows Confederate troop areas, tents, gun emplacements and where the orders were dropped. On the lower right, it says: **Lost pouch left here**.

The packet was dropped away from the Officer's tents in the hitching area. Previously, where Hill's campground was located and the orders were lost, has been in dispute.

After the war, Daniel Harvey Hill strenuously denied anyone having authority to sign for the orders ever received the packet. Hill received his instructions from Stonewall Jackson's command and was not in need of a copy from Lee's Headquarters. True, but telling the Theater Commander you do not need his orders would be very unusual. Receiving orders directly from the top represented upward mobility in the command structure which Hill certainly would have wanted.

Evidential Details

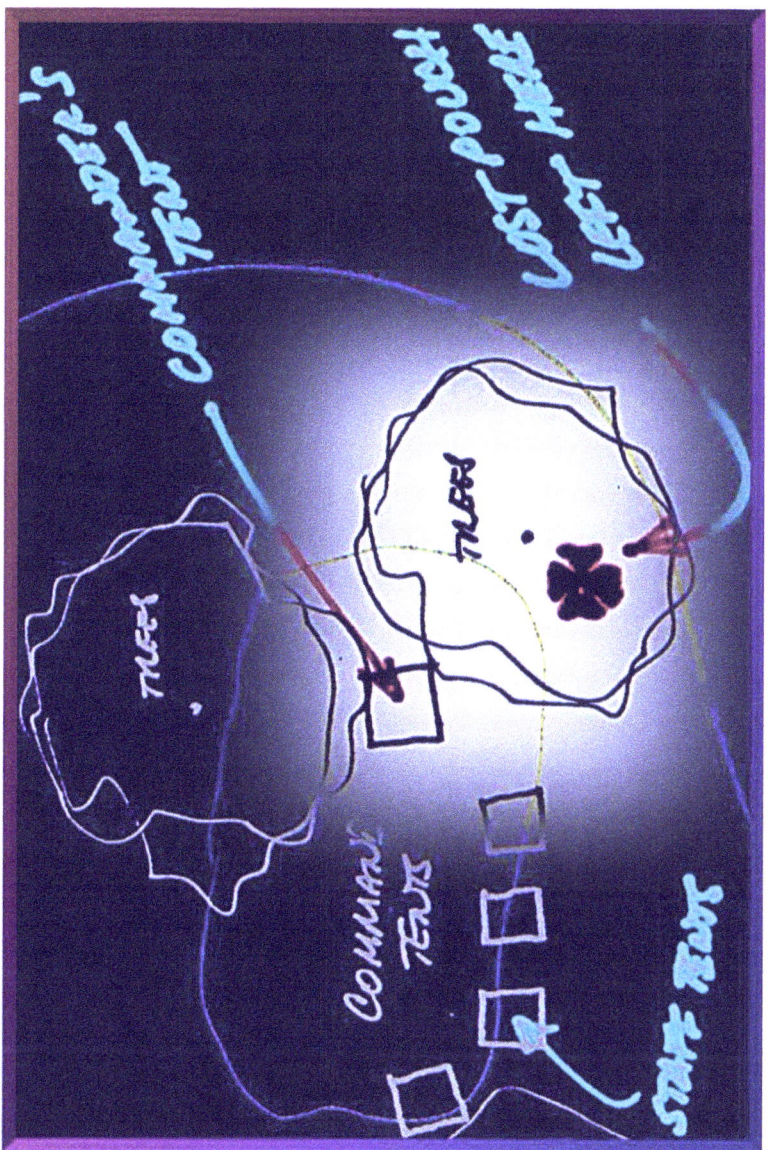

Confederate Command Tent Close-up

In the military, McMoneagle was tasked with how and where various clandestine transactions took place. This picture details the distance between the Command tents and where the packet was dropped.

The Lost Order

Politically imaging is important to any General with a new direct report. What Robert E. Lee thought mattered and something was not right here. Whether a staff orderly had technically signed for the orders was not Lee's bottom line. Top secret order of battle plans had been mishandled putting the entire Southern war effort in jeopardy. It would make sense that whoever lost the orders had a second copy. How else could he operate over a week without an inquiry to headquarters about not receiving orders? So how did the packet arrive?

Colonel Archer Anderson, D.H. Hill's Chief of Staff, received the orders packet from courier Henry Kyd Douglas.

McMoneagle – It was read to him by his Chief of Staff, a colonel, who got it from a courier [rider].

So the orders were read to him by Colonel Archer Anderson. This confirms Hill's adjutant Rachford had not lied when he signed the affidavit stating he never received the orders. It is just that Hill had the wrong man swear. We could find no documentation that either Officer ever mentioned this event.

The insider probabilities were that Lee's courier swore he delivered the orders to Colonel in Hill's camp, and General Jackson acknowledged he had sent Hill another copy. Faced with this technically unsubstantiated innuendo, Hill wrote his old adversary George McClellan a letter dated April 17, 1869. In it, he stated that all Confederate orders packets were returned to headquarters after the orders were removed. The fact the orders packet was found intact proved Hill was innocent.

But, as could be imagined, old adversary McClellan was not in a position to verify Confederate Staff Officer paperwork flow

Evidential Details

details. It does however show the length Hill was prepared to go to win support for a cause that no longer mattered excepting pride and vindication. But, if there is a "smoking gun" of doubt, it would be to understand Hill's need to proclaim that the orders loss actually saved the day!

<u>D.H. Hill</u>: *"In the battle of South Mountain the imaginary forces of the Lost Dispatch were worth more to us than ten thousand men."*[lvii] *"The ten imaginary regiments of the Lost Dispatch retarded his* (General John Gibbon's [1827-1896]) *progress through the woods..." "There were only a few skirmishers on his right, but the lost dispatch made him believe otherwise." "The loosing of the dispatch was the saving of Lee's army."*[lviii]

Hill wrote this even though the lost orders did not mention regiments or numbers, and Gibbon did not have a copy! Curiously, Hill would be the only Confederate to make positive state-ments about the packet's loss. But why sanctify the South's worst security breach that resulted in having to scramble to avoid disaster? And then to go into a congratulatory mode about it being the Army of Northern Virginia's saving grace was simply beyond belief.

This agitated Robert E. Lee who was then Washington College President. In uncharacteristic manner, on February 15, 1868, he indignantly condemned Hill's remarks to a college clerk named E.C. Gordon who jotted his comments down. Later that day Lee complained to Professor William Allan, who went on to become an Army of Northern Virginia historian.

Still grousing a week later, Lee wrote D.H. Hill about the lost orders on February 21, saying:

"I consider it a great calamity" and that, *"subsequent reflection has not caused me to change my opinion."*[lix]

McClellan's delayed troop movements allowed Harpers Ferry to be captured. It also allowed the South to regroup behind Antietam Creek enabling Lee's 38,000 effectives to fight the 87,000 Federal's[lx] to a tactical draw. What should have been an easy victory remains the single bloodiest day in American history.

The Lost Order

As the sun rose the next morning, the Army of Northern Virginia, now about 28,500 strong, started moving their wounded south to the Potomac River, while waiting for an attack from McClellan's remaining 74,500 men. None came. As the Confederate army slipped away, 32,000 Union soldiers had not participated in what became a strategic victory for the North.

McMoneagle – Southern retreat - I think the (Potomac) **river crossing is important as I get a sense that the Southern Army was allowed to retreat across it, when all the Union had to do was pin them with their backs to the river and the war might have ended long before it even really got started. Such a move most certainly would have eradicated nearly an entire army of the South.**

For his failure to press the enemy, George McClellan was again relieved of command. In the 1864 elections he was the Democratic Presidential candidate running against Abraham Lincoln. In that campaign, the Democratic Party Platform called for an immediate cease fire while sizable Confederate armies remained in the field.

So in September of 1862, Confederate arms were checked in the State of Maryland northwest of Washington. Lee had lost control of events and was unable to fight at his chosen place with his forces aligned as he would have preferred. The finding of the lost orders made the difference in battle location, the campaign's outcome, and hence, the Emancipation Proclamation's timing.

Overall, the Maryland Campaign resulted in 27,475 Union losses including those captured at Harpers Ferry and the Battles at South Mountain. Confederate losses were reported at 12,392. Assuming 66 more men on each side died or were taken captive, a 40,000 man rounded combined total of dead, missing, wounded or captured, between September 12 and 17 is reasonable. The removal of 40,000 men, between the ages of 16 and 36, would comprise approximately 22.5% of a city with 175,000 inhabitants. In just five days, every male in this age bracket disappeared. Such

Evidential Details

was the American Civil War's devastation.

How *Special Order 191* was lost has always been the most important mystery in America's Civil War. And its significance was even more pronounced when European governments declined to extend diplomatic recognition to Richmond's commercial attaché offices. While sympathetic to the South for their need of quality cotton and tobacco, they finalized their decision based on the outcome at Antietam Creek and the issue of slavery

As the Illinois Republican Senatorial candidate in 1858, Lincoln had established his pro-emancipation credentials well. So much so that, reinterpretationists aside, his 1860 Presidential victory was a signal to the South to take up secession. But Lincoln waited to issue the proclamation until he had some success on the battlefield. After Antietam, Abraham Lincoln came forward and told his Secretaries in a cabinet meeting that, *"God has decided this question in favor of the slaves."*[lxi] Within eighteen months, he had transformed the war into a dual cause to abolish slavery as well as preserve the Union.

This nation saw the results when, on November 4, 2008, one legacy of the lost order was realized. For it was on that day Barak Hussein Obama was elected President of the United States. And so, "The course of the war and the course of the nation were forever changed as a result."[lxii]

Aftermath

The direct military results of the lost order were the Battles of South Mountain and Antietam Creek. The day after the Antietam battle, Lee held his ground while the army evacuated its trains of wounded across the Potomac River at Shepherdstown, Virginia.

The 12,419 Union prisoners captured at Harpers Ferry were not sent south. Instead, they were paroled by raising their right hands and swearing an oath in unison not to serve against the Confederacy until such time as an equal number of Southern

The Lost Order

men were paroled. Each officer had to sign a statement.

The Union parolees started their two day march to Frederick, M.D. on the 16th, and going to Annapolis, Maryland. The Army decided not to furlough the men instead railing them to Chicago's Camp Douglas on September 23rd. Their final destination was Minnesota where they were to fight the Sioux Indians under General Pope's command. However, they were never sent because most of the officers were called to Washington to testify at the Harpers Ferry Military Court of Inquiry.

On September 25, 1862, the Secretary of War (Defense) opened a military commission, "*to investigate the circumstances of the abandonment of Maryland Heights*" which led directly to Harper's Ferry's surrender. Since commanding officer Dixon S. Miles had been killed with what seemed was the last artillery shell blast, the ax fell on Colonel Thomas H. Ford of the 32nd Ohio.

On November 8, he was, "*dismissed from the service of the United States*", for having conducted his Maryland Heights defense, "*without ability, abandoned his position without sufficient cause, and has shown throughout such a lack of military capacity as to disqualify him, in the estimation of the Commission, for a command in the service.*"[lxiii] All the Harpers Ferry prisoners were declared exchanged by January 10, 1863.

On October 3, President Lincoln traveled to the Antietam battlefield to confer with General McClellan. Questions arose as to why the retreating Confederate Army had not been pressed against the Potomac River. McClellan's commitment to the Union cause has subsequently come into question and this may have been in the back of the President's mind as they conferred.

On July 7, 1862, just two months before the finding of the lost order, McClellan had handed President Lincoln a letter at Harrison's Landing in Virginia. The letter said in part that Federal: "*Military power should not be allowed to interfere with the relations of servitude, either by supporting or impairing the authority of the master...*"[lxiv] This mirrored the Confederate Congress's position.

Evidential Details

During that summer, President Lincoln had been working on his Emancipation Proclamation. On July 22, 1862, it was read to his cabinet. Now, due to the Lost Order, the President had the victory he needed to issue his proclamation. The new law read in part:

That on the 1st day of January A.D. 1863, all persons held as slaves within any State or designated part of a State...shall be then, thence-forward, and forever free...

And I further declare and make known that such persons of suitable condition will be received into the armed service of the United States to garrison forts, positions, stations, and other places, and to man vessels of all sorts in said service.

And upon this act, sincerely believed to be an act of justice, warranted by the Constitution...I invoke the considerate judgment of mankind and the gracious favor of Almighty God.

Addendum

These are Confederate General Thomas J. Jackson's rewritten orders to General Daniel Harvey Hill. Here the historian can review the top secret orders Union General McClellan had in his possession as the Confederate Army was split five ways.

Special Orders, No. 191
HEADQUARTERS, ARMY OF NORTHERN VIRGINIA
September 9th, 1862

The Army will resume its march tomorrow, taking the Hagerstown road. General Jackson's command will form the advance, and after passing Middletown, with such portions as he may select, take the route toward Sharpsburg, cross the Potomac at the most convenient point, and by Friday night take possession of the Baltimore and Ohio Railroad, capture such of the enemy as may be at Martinsburg, and intercept such as may attempt to escape from Harpers Ferry.

The Lost Order

General Longstreet's command will pursue the same road as far as Boonsboro', where it will halt with the reserve, supply, and baggage trains of the army.

General McLaws with his own division and that of General R.H. Anderson, will follow General Longstreet; on reaching Middletown he will take the route to Harpers Ferry, and by Friday morning possess himself of the Maryland Heights and endeavor to capture the enemy at Harpers Ferry and vicinity.

General Walker, with his division after accomplishing the object in which he is now engaged, will cross the Potomac at Check's ford, ascend its right bank to Lovettsville, take possession of Loudoun Heights, if practicable, by Friday morning, Keyes's ford on his left, and the road between the end of the mountain and the Potomac on his right. He will, as far as practicable, cooperate with General McLaws and General Jackson in intercepting the retreat of the enemy.

General D.H. Hill's division will form the rearguard of the army, pursuing the road taken by the main body. The reserve artillery, ordnance, and supply trains, etc., will precede General Hill.

General Stuart will detach a squadron of cavalry to accompany the commands of Generals Longstreet, Jackson, and McLaws, and, with the main body of the cavalry, will cover the route of the army and bring up all stragglers that may have been left behind.

The commands of Generals Jackson, McLaws, and Walker, after accomplishing the objects for which they have been detached, will join the main body of the army at Boonsboro' or Hagerstown.

Each regiment of the march will habitually carry its axes in the regimental ordinance-wagons, for use of the men at their encampments, to procure wood, etc.

By command of General R.E. Lee
R. H. Chilton, Assistant Adjutant-General

Evidential Details

The Lost Order

The Lost Order

[i] Hill, Daniel Harvey, Lieutenant General, C.S.A., *"The Battle of South Mountain, or Boonsboro"*; *Battles And Leaders of the Civil War, Vol. II 1887*; p. 565, hereafter referred to as B&L
[ii] *The Official Records of the War of the Rebellion*, Vol. XIX, Part ii, pp. 590-603, hereafter referred to as OR
[iii] Hood, J.B., Lt. Gen, CSA, *Advance and Retreat*; Blue and Gray Press a division of Book Sales, Inc. 1985; p. 41
[iv] Davis, Burke, *Jeb Stuart – The Last Cavalier*; Wings Books, 1957; p. 191
[v] ibid; p. 192
[vi] B&L; [vi] Hill, Daniel Harvey, Lieutenant General, C.S.A., *The Battle of South Mountain, Fighting For Time at Turner's and Fox's Gaps*; p. 560
[vii] Davis; p. 189
[viii] *McClellan Papers {C-7:63}*; Library of Congress; United States of America
[ix] Sears, Stephen W., *George B. McClellan: The Young Napoleon*; Ticknor & Fields 1988; p. 276; footnote p. 443
[x] Farwell, Byron, *Stonewall - A Biography of General Thomas J. Jackson*; W.W. Norton & Company, 1993, p. 428
[xi] ibid; p. 428 - 9
[xii] Brown, Edmund R.; *The Twenty-seventh Indiana volunteer infantry in the war of the rebellion, 1861 to 1865, First division, 12th and 20th corps by a member of Company C.*; 1899; all four quotes from p. 226-228
[xiii] Sears, Stephan W., *Landscape Turned Red*; Popular Library, Warner Books, Inc., 1983 p. 123, hereafter referred to as LTR
[xiv] Fishel, Edwin C., *The Secret War for the Union - The Untold Story of Military Intelligence in the Civil War*; Houghton Mifflin Company, 1996, p. 222
[xv] Bowers, John, *Stonewall Jackson - Portrait of a Soldier*; William Morrow & Co. Inc. 1989; p. 299
[xvi] Sears, Stephen W., editor, *The Civil War Papers of George B. McClellan*; Ticknor & Fields; 1989, p. 434
[xvii] Colgrove, Silas, Brevet Brigadier-General, U.S.V., *The Finding of Lee's Order*; B&L Vol. II; p. 603 – all three quotations.
[xviii] B&L Vol. II; p. 603
[xix] LTR; p. 124
[xx] ibid; p. 603
[xxi] ibid
[xxii] Freeman, Douglas Southhall, *Lee's Lieutenants* Vol. II; Charles Scribner & Son's 1948, p. 718
[xxiii] Sears; p. 282
[xxiv] Frye, Dennis E., *The General's Tour – Stonewall Attacks – The Siege of Harpers Ferry*; Blue and Gray Magazine; August-September 1987; p. 18
[xxv] LTR; p. 385
[xxvi] Sears; p. 280 & 282
[xxvii] Fishel; p. 239
[xxviii] Fishel; p. 240
[xxix] Fishel; p. 238
[xxx] B&L Vol. II; p. 560
[xxxi] Huyette, Miles Clayton, *The Maryland Campaign and the Battle of Antietam*; Buffalo, N.Y. 1915, p. 21
[xxxii] B&L Vol. II; p. 564-565
[xxxiii] ibid; p. 565
[xxxiv] ibid; p. 563
[xxxv] ibid; p. 566

Evidential Details

[xxxvi] ibid; p. 566-567
[xxxvii] ibid; p. 570
[xxxviii] Sears; p. 285
[xxxix] B&L; p.594 - William B. Franklin, Major General United States Volunteers, *Notes on Crampton's Gap and Antietam*
[xl] Howard, H.E., *The Charlottesville*, article entitled *Lee, Lynchburg, and Johnson's Bedford Artillery*; Lynchburg, Virginia, 1990; p.139
[xli] *LTR*, p.159
[xlii] *B&L Vol. II*; p. 580
[xliii] ibid; p. 605 - John G. Walker, Major General, C.S.A., *Jackson's Capture of Harpers Ferry*
[xliv] Bridges, Hal, *Lee's Maverick General*, McGraw Hill Publishing; 1961; p. 96
[xlv] *LTR*; Appendix I; *The Lost Order*; p.381
[xlvi] *Fishel*, p. 223; "For Lee's comment, see Murfin, *Gleam of Bayonets*, p 337
[xlvii] *B&L Vol. II*; p. 559
[xlviii] Wert, Jeffery D., *General James Longstreet - The Confederacy's Most Controversial General. A Biography*; Simon and Schuster, 1993, p. 93
[xlix] Haskell, John Cleves, *The Haskell Memoirs*; G.P. Putnam's Sons; 1960; p.45
[l] *Bridges*; Longstreet's statement p. 149
[li] *B&L Vol. II*; p.560
[lii] *Longstreet Papers*, Duke University Library; D.H. Hill's letter to James Longstreet, Feb. 11, 1885
[liii] Symonds, Craig L., *Joseph E. Johnston - A Civil War Biography;* W.W. Norton, 1992, p. 142
[liv] *Daniel Harvey Hill, Jr. Papers*, North Carolina Division of Archives and History, Raleigh, N.C.; Letter from C.D. Fishburne to D.H. Hill, Jr., Chancellorsville, VA. Feb. 8, 1890; p. 1
[lv] *Bridges*; p. 6
[lvi] Wilmington Messenger, September 27, 1889
[lvii] ibid; p. 573
[lviii] ibid; p. 570
[lix] *Lee Papers*, Library of Congress, Lee to D.H. Hill, February 21, 1868.
[lx] *B&L Vol. II*; 'The Opposing Forces in the Maryland Campaign'; p. 603
[lxi] *Sears*; p. 325
[lxii] Sears, Stephen W., *The Civil War Battlefield Guide* "Antietam Chapter"; Houghton Mifflin Company; 1990; p. 86
[lxiii] Military Commission appointed September 23, 1862; presided over by Major General David Hunter pursuant to War Department General's Office Order 183; *Conclusionary statement* signed by order of the Secretary of War, Assistant Adjutant-General E. D. Townsend, dated Washington D.C. November 8, 1862; Library of Congress
[lxiii] LS, Lincoln Papers, Library of Congress. OR, Ser. 1, XI, Part 1, pp.73-74

The Lost Order

Bibliography

- Battles and Leaders of the Civil War, Vol. II. Castle 1887
  --- Hill, Daniel Harvey, Lieutenant General, C.S.A., *The Battle of South Mountain, or Boonsboro*
  --- Hill, Daniel Harvey, Lieutenant General, C.S.A., *The Battle of South Mountain, Fighting For Time at Turner's and Fox's Gap*
  --- Colgrove, Silas, Brevet Brigadier-General, U.S.V., *The Finding of Lee's Order*
  --- William B. Franklin, Major General United States Volunteers, *Notes on Crampton's Gap and Antietam*
  --- John G. Walker, Major General, C.S.A., *Jackson's Capture of Harpers Ferry*
  --- Editors, The Opposing Forces in the Maryland Campaign
- Bowers, John, *Stonewall Jackson - Portrait of a Soldier*; William Morrow & Co. Inc. 1989
- Bridges, Hal, *Lee's Maverick General*, McGraw Hill Publishing; 1961
- Brown, Edmund R.; *The Twenty-seventh Indiana volunteer infantry in the war of the rebellion, 1861 to 1865, First division, 12th and 20th corps, by a member of Company C*. 1899
- *Daniel Harvey Hill, Jr. Papers*, North Carolina Division of Archives and History, Raleigh, N.C. 1890
- Davis, Burke, *Jeb Stuart – The Last Cavalier*; Wings Books, 1957
- Farwell, Byron, *Stonewall - A Biography of General Thomas J. Jackson*; W.W. Norton & Company, 1993
- Fishel, Edwin C., *The Secret War for the Union - The Untold Story of Military Intelligence in the Civil War*; Houghton Mifflin Company; 1996
- Freeman, Douglas Southall, *Lee's Lieutenants* Vol. II; Charles Scribner & Son's; 1948
- Haskell, John Cleves, *The Haskell Memoirs*; G.P. Putnam's Sons; 1960
- Hood, J.B., Lt. Gen, CSA, *Advance and Retreat*; Blue and Gray Press, a division of Book Sales, Inc.; 1985
- *Lee Papers*, Library of Congress; February 21, 1868
- *The Longstreet Papers*, Duke University Library
- Library of Congress; United States of America
  --- *McClellan Papers {C-7:63}*
  --- War Department Office Order 183; Conclusionary statement Assistant Adjutant-General E.D. Townsend, Washington D.C. November 8, 1862
- Sears, Stephen W., *George B. McClellan: The Young Napoleon*; Ticknor & Fields; 1988
  --- *Landscape Turned Red*; Popular Library, Warner Books, Inc.; 1983
  --- *The Civil War Papers of George B. McClellan*; Ticknor & Fields; 1989
  --- *The Civil War Battlefield Guide* "Antietam Chapter"; Houghton Mifflin

Evidential Details

    Company; 1990
- Symonds, Craig L., *Joseph E. Johnston - A Civil War Biography;* W.W. Norton; 1992
- War of the Rebellion: The Official Records of the Union and Confederate Armies, Vol. XIX, Part ii, U.S. Government Printing Office 1880
- Wert, Jeffery D., *General James Longstreet - The Confederacy's Most Controversial General. A Biography*; Simon and Schuster; 1993

Magazines – Newspapers

--- Frye, Dennis E., *The General's Tour – Stonewall Attacks – The Siege of Harpers Ferry*; Blue and Gray Magazine; August-September 1987
--- Howard, H.E., The Charlottesville, article entitled *Lee, Lynchburg, and Johnson's Bedford Artillery*; Lynchburg, Virginia
--- Wilmington Messenger, September 27, 1889

The further we move away from the present and into the past, the more likely changes in that history have taken place, the more likely what we think we are seeing as real, isn't.

...the past is a chameleon that always wears a tint of the "now." It fools us into thinking it is, or always was an absolute, when, in fact, it has never been that way.

        Both quotes Joseph McMoneagle - *The Ultimate Time Machine*

Evidential Details

(The Planet Jupiter targeting) *Experiment #46 lay obscure between 1974 and 1979. No continuing attempt was made to feedback other of its categories, and the SRI* (Stanford Research Institute) *work progressed along more immediately fruitful* (feedback) *lines.* (Six years later) *The 1979* (Voyager II) *scientific discovery and* (the July 9, 1979) *confirmation of* (Jupiter's) *the Jovian Ring came as one of the larger shocks - and surprises -- in astronomical history.*

The entirety of the Jupiter Probe raw data was now organized and compared to scientific feedback -- after which all of the data, except the mountains, could be seen as near-approximately confirmed. Now, however, the formal report was generally rejected. Yet word got around.

THE 1973 REMOTE VIEWING PROBE OF THE PLANET JUPITER
Military Intelligence Remote Viewing Protocol developer Ingo Swann - (12Dec95)

...there is also a tendency on the part of almost everyone I've seen running a study to have a successful run and then say, "Hey! That gives me an idea! I'll bet that if we _____, it would go even better!!!" From there on, you're just shooting from the hip.

Professionalism includes getting a clean protocol and sticking to it throughout the whole process, no matter what great idea(s) you get. New ideas and changes should be tested on their own at a different time and in a different study.

Star Gate Data Base Manager Lynn Buchanan e-mail November 08, 1999

The ease with which viewers can move their minds through time is one of the major strengths of Controlled Remote Viewing. In CRV, no distinction is made between time and space as far as any working conditions are concerned. It is as easy in CRV to move back ten days as it is to move back ten feet.

Naturally psychics love to build castles and give explanations for what they perceive. When not allowed to do so, they tend to get a little testy, but also produce much higher quality information. In the end, that has to be the bottom line.

Both quotes from Lyn Buchanan in *The Seventh Sense*

Evidential Details

Part III

...as a result of my own previous exposure to this (remote viewing) community I became persuaded that war can almost always be traced to a failure in intelligence, and that therefore the strongest weapon for peace is good intelligence.

~ H. E. Puthoff, PhD. ~

Founder and First Director (1972-1985)
The Military Intelligence program known as Operation Star Gate

Evidential Details

Credentials

JOSEPH W. MCMONEAGLE, CW2, US Army, Ret., KCStS
Owner/Executive Director of
Intuitive Intelligence Applications, Inc.

Mr. McMoneagle has over 45 years of professional expertise in research and development, in numerous multi-level technical systems, the paranormal, and the social sciences. Experience includes: experimental protocol design, collection and evaluation of statistical information, prototype design and testing, Automatic Data Processing equipment and technology interface, management, and data systems analysis for mainframe, mini-mainframe, and desktop computer systems supporting information collection and analysis for intelligence purposes.

He is currently owner and Executive Director of Intuitive Intelligence Applications, Inc., which has provided support to multiple research facilities and corporations with a full range of collection applications using Anomalous Cognition (AC) in the production of original and cutting edge information. He is a full time Research Associate with The Laboratories for Fundamental Research, Cognitive Sciences Laboratory, Palo Alto, California, where he has provided consulting support to research and development in remote viewing for 16+ years. As a consultant to SRI-International and Science Applications International Corporation, Inc. from 1984 through 1995, he participated in protocol design, statistical information collection, R&D evaluations, as well as thousands of remote viewing trials in support of both experimental research and active intelligence operations for what is now known as Project STARGATE. He is well versed with developmental theory, methods of application, and current training technologies for remote viewing, as currently applied under strict laboratory controls and oversight.

During his career, Mr. McMoneagle has provided professional intelligence and creative/innovative informational support to the Central Intelligence Agency, Defense Intelligence

Credentials

Agency, National Security Agency, Drug Enforcement Agency, Secret Service, Federal Bureau of Investigation, United States Customs, the National Security Council, most major commands within the Department of Defense, and hundreds of other individuals, companies, and corporations. He is the only one who has successfully demonstrated his ability more than two dozen times, by doing a live remote viewing, double-blind and under controls while on-camera for national networks/labs in four countries.

Mr. McMoneagle has also been responsible for his Military Occupational Specialty at Army Headquarters level, to include control and management of both manned and unmanned sites within the Continental United States, and overseas. He was responsible for all tactical and strategic equipment tasking, including aircraft and vehicles, development of new and current technology, planning, support and maintenance, funding, training, and personnel. He has performed responsibly in international and intra-service negotiations and agreements in support of six national level intelligence agencies, and has acted as a direct consultant to the Commanding General, United States Army Intelligence and Security Command (INSCOM), Washington D.C., as well as the Army Chief of Staff for Intelligence (ACSI), Pentagon.

"For the intelligence world, mental access of target people has great value. Profilers and psychological analysts who specialize in this craft are highly prized within the intelligence community. They are employed in that part of the intelligence community which is called HUMINT, which is short for "Human Intelligence," or "intelligence derived from human sources."

Lyn Buchanan – The Seventh Sense

Evidential Details

Human Use

Remote Viewing research sometimes involves input from different sources as in the application of the Army's Human Use Policies developed to protect soldiers after the accidental deaths in their LSD investigation.

"In February 1979, the General Counsel, the Army's top lawyer, declared [the RV Program named] Grill Flame activities constitute Human Use." The Unit, "… was in the middle of the [authorization] process in March 1979 when the Human Use determination was reversed by the Army Surgeon General's Human Use Subjects Research Review Board. Their decision…trumped the Army General Counsel's ruling…" "On November 20, the Surgeon General's board changed its mind and decided that Grill Flame did indeed involve Human Use. It took until February 1, 1982 to get final approval from the Secretary of the Army to continue operations." [1]

New candidates were then issued a warning by a Major General before being accepted into the super secret 902nd Intelligence Unit.

"Among other things, they noted that if he joined the project, he would be exposed to psychic phenomena at a level and with a frequency that most people had never experienced before. As a result, he might change in certain ways. Ultimately, no harm should come to him, but he might have a new perspective on himself, his marriage, the universe. In a sense, he might become a new man, and a new husband."

The candidate and his wife were advised to talk, "…this over before they made the final commitment to go to Fort Meade." [2]

---

[1] Smith, Paul H., *Reading the Enemy's Mind – Inside Star Gate, America's Psychic Espionage Program*; Tor Non-fiction, 2005; p. 118

[2] Schnabel, Jim, *Remote Viewers: The Secret History of America's Psychic Spies*; Dell Non-Fiction, 1997, p. 270

Evidential Details
A CHINESE ENCOUNTER

The United States is not the only nation to study and use Remote Viewing. Below is a story allowing enthusiasts and skeptics alike a rare look at life inside the Unit during the middle 1980's.

The first time it happened was right after [Major] General Stubblebine had briefed me on the project and said that I would be contacted. The next week I was working mid shift, and one of the afternoons, I lay down for a nap. In that nap, I had a really shallow and lame dream about something I can't remember now. But at one point, right over the top of that dream there was what appeared to be a semi-translucent visual of three people.

One was a very respectable, businesslike slender man in a suit. A second was a very burly, stocky man, also in a suit, and with a very "Texas farmer" face. The third was an...oriental girl... (I find it impossible to tell the age of oriental women). She was following along behind the two men and watching.

The men came up to me and talked about something, but I couldn't hear them. The girl was standing behind the two men, listening. The faces were very clear. Clear enough that when the two men actually came to [the INSCOM[1] Base in] Augsburg [Germany] to interview me, I recognized them immediately. I could have picked them out of a crowd on the sidewalk. I didn't think anything of the fact that the girl wasn't with them. It would have been odd to have her on a military trip overseas. I thought she was probably someone in the unit.

Months later, when I got to the unit, she wasn't there. I asked about her and neither the director nor Joe [McMoneagle] (the two men who came to interview me) knew who I was talking about. I figured that it was just an AOL (STRAY CAT),[2] and blew it off.

About a year later, I was doing a practice target. The target was a museum at Arizona State University (I didn't know that - I

---

[1] INSCOM is the abbreviation for the Army's Intelligence and Security Command.

[2] Stray Cat is a viewer acronym describing the Subconscious Transfer of Recollections, Anxieties, and Yearnings to Consciously Accessible Thought.

Chinese Encounter

only had numbers). I was describing things lying in glass topped cases, with the cases up on legs and stands, all arranged around the room for easy access, when I noticed that someone at the target site was looking straight at me, as though she could see me. It startled me, and for probably the only time ever, I wasn't startled OUT of the session, but deeper into it. I looked back at her, and realized that it was the same girl who had been following the director and Joe in my earlier "dream", back in Augsburg. I looked directly at her, and started to say hello, but then she realized that I could see her, too, and she half turned, and disappeared. That threw me out of the session.

Fortunately, [Captain] Paul Smith was my monitor, and ever the curious one, when I told him what had happened, he said, "Let's follow her and see where she went." Through a series of very impromptu movement commands, we finally located her back at the place where she worked ... the Chinese psychic intelligence effort.

She appeared in some of my sessions after that, but rarely. I tried to find her several times, and a few of them succeeded. Apparently, what they defined as "session" and what we defined as "session" weren't the same. Anyway, over time, we struck up somewhat of a stand-offish acquaintance. About a year after that, I hadn't bumped into her again, so I did a session specifically to find her. She was then in college in a very large city, and evidently out of the government's project altogether. When I found her, she acknowledged my presence, and very strongly desired that we not have further contact. I backed out of the session, and haven't tried again, since. Don't cha love war stories?" [3]

    Oct. 1, 1998 e-mail from Leonard Buchanan – Former Operational Database Manager 902nd Military Intelligence Unit - Fort Meade, Maryland and Owner of Problems>Solutions>Innovations, Inc.

[3] For more information, see, *China's Super Psychics* by Paul Dong and Thomas Raffill; Marlowe & Co. New York, 1997

Evidential Details

Remote Viewing Protocols

Surrounding the military's RV session protocols are the Operational Flow Protocols. The tasking agency was the "Customer" whose identity was strictly withheld to avoid inferences leading to Analytic Overlay. First published here, this process was highly classified for over two decades.

* * *

"In actual fact, there was pretty much a different work set-up every time we changed directors in the military unit which was pretty often as projects go. As a result, the "ideal plan" was never adhered to. Many times, we had to sort of switch horse in mid-stream. Anyway, here is the "ideal" workflow:

The **CUSTOMER** (Governmental Agency) comes to the unit director with a tasking.
The **UNIT DIRECTOR** meets with the customer and:
1) makes absolutely certain that the customer knows what CRV is and isn't – what it will and won't do.
2) looks the customer's problem over to see that it is the type of work we are best suited for. If not, he suggests a different solution for them.
If so, he then:
3) gets rid of the customer's "test" questions which only take up time and effort and accomplish nothing.
4) gets rid of the unnecessary questions – just fluff questions which the customer would like to have answered.
5) makes certain the questions asked are questions the customer really wants the answers to. There are LOTS of times when the customer will ask, "Who killed the victim", when the information he really wants is, "Where can we find the evidence that will show who killed the victim?"
6) agrees in writing on a set of basic questions which will be answered, once all the fluff and confusion is gotten out of the way.
7) makes certain that the Customer knows that these questions will be answered, and that other information will be provided, if it is found. However, if it isn't found, then the viewers are only responsible for what is being tasked. Follow-on questions will have

Protocols

to be asked later.
8) explains to the Customer the need for accurate feedback.
9) gets a definite commitment from the Customer that such feedback will be given, on each and every viewer's answer(s) to each and every question.
10) sets a commitment date for providing the answers. This must be a realistic date. Every Customer wants answers right now or yesterday, but the unit director needs to impress on the Customer that there are other customers who also have time limits of now or yesterday, and that reality must figure into the planning, like it or not.
11) provides the final list of questions to the Project Officer, along with any background information about the case gained from the customer.

The **PROJECT OFFICER** studies the background information and tasked questions and:
1) determines the main subject matter for each question.
2) decides the project number and fills out all the preliminary paperwork required for starting a new project.
3) provides the list of subjects to the Data Base Manager. The Data Base Manager looks up each information category in the data base and provides the Project Manager with a separate list of Viewers' names as suggested Viewers for each question.
4) determines which Viewers and Monitors should work on each question.
5) looks at the Viewers' and Monitors' existing schedules and determines the project's time line. He may even do a Pert chart to make scheduling easier.
6) "translates" each question into neutral wording.
7) notifies each Monitor and Viewer of the work schedule change.
8) generates an official tasking sheet to hand to each Monitor.

The **MONITOR** receives the tasking and coordinates from the Project Officer, along with any background information the Project Officer thinks the Monitor should know to help the Viewer better perform a productive session. The Monitor then:
1) makes certain he knows the Viewer's likes and dislikes, quirks, micro-movements, etc. If not, these are either looked up or found out from another Monitor who is more familiar with the Viewer.

Evidential Details

2) gets information from the Database Manager about the Viewer's strengths and weaknesses. While this carries the danger of a "self-fulfilling prophecy", the Monitor is hopefully trained enough to use the information for formatting the session, rather than for guiding and leading the Viewer. If the Monitor is not this well trained, this step is passed up.
3) prepares the session workplace.
4) goes through the session with the Viewer.
5) helps the Viewer write the summary, if necessary.
6) after the paperwork is all done, provides both the Viewer's transcript and his (the Monitor's) session notes to the Analyst.

The **ANALYST** receives the paperwork and:
1) familiarizes himself with all the background knowledge.
2) collects the papers from all Viewer/Monitor pairs.
3) looks into his own notes on each and every Viewer to see work profiles (prone to using imagery, prone to using allegories, etc.). The Database Manager can be of help in this step.
4) performs analysis on the session (see the Analyst's Manual).
5) writes up his reports, critiques, summaries, etc. and provides it to the Report Writer.

The **REPORT WRITER** receives all the information from the Analyst and:
1) familiarizes himself with all the available background information.
2) familiarizes himself with all the Analyst's finding, interpretations and comments.
3) writes the final report (see the Report Writer's Manual)
NOTE!!! This includes taking the finalized answer to each Viewer to make certain that what is being reported is what the Viewer actually meant to say.
4) provides the final report to the Project Officer.

The **PROJECT OFFICER** then:
1) receives the finalized answers to each question after the session has been performed, analyzed and prepared for reporting.
2) gives final approval on the final report.
3) passes the final report to the Unit Director for delivery to the Customer.

The **UNIT DIRECTOR** then:

Protocols

1) contacts the Customer and sets a date and time to go over the report. Information is not given ad hoc over the phone, nor is an "executive summary" provided.
2) meets with the Customer to provide the report.
3) once again makes certain that the Customer understands the CRV process, strengths and limitations.
4) explains what happened, and how each answer was obtained.
5) points out to the Customer that each question has a "dependability rating" beside it which will tell the Customer what each Viewer's track record is on each specific answer to each type of question. He explains how this "dependability rating" can be used by the Customer as an aid to making decisions from the information provided.
6) sets – in writing – a hard and definite "drop dead" date for feedback.
7) if/when feedback comes in, provides it to the Project Officer who handled the case.
8) if feedback doesn't come in, or is received incorrectly, it is returned to the Customer to either, "dun him" for feedback, or to re-explain how feed-back needs to be provided, formatted, etc.

The **PROJECT OFFICER** then:

1) evaluates each Viewer's response to each question against the feedback.
2) provides an evaluation to each Viewer.
3) provides accurate data to the Database Manager for input into the database.
4) completes all summary paperwork for the project.
5) organizes all related paperwork, checks it for completeness, and prepares it for final storage and filing.

The **DATABASE MANAGER**:

1) inputs all received information into the database.
2) "massages" the database to provide information to those who need it. This includes the Training Officer and all Trainers.
3) maintains quality control on the data going in. "Garbage in – garbage out".

The **TRAINING OFFICER**:

1) schedules training times and facilities.
2) keeps evaluation reports on the Trainers.

Evidential Details

The **TRAINER**:
1) accompanies new Viewers through the training process, analyzing their needs and progress every step of the way (see Trainers Manual).
2) makes and keeps records of the Viewer Student's "natural micro-movements". These will be provided to the Monitors along with a Viewer Student's profile of strengths and weakness.
3) advises management of the Viewer Student's progress and advises as to the student's best possible "training track" for providing the most useful and productive Viewer possible.

Needless to say, this is an overview, and not a complete list of responsibilities and obligations. For example, it doesn't cover what goes on in follow-on tasking, etc.

July 23, 1998 e-mail from: Leonard Buchanan– Former Operational Database Manager at the 902nd Military Intelligence Unit - Fort Meade, Maryland and Owner of Problems> Solutions>Innovations, Inc.

Interview Clarification

Question: Generally speaking, how much...information should be given a viewer in operations / applications?

Joseph– McMoneagle: None. Zero. What you can do if the target requires a response or a description of an individual, you can say, "*Describe the individual at* (whatever location)" and the location needs to be hidden (would be a number, for instance). If you were targeting let's say a church, and there was an individual in that church, the church would be coded as say, "location A1". It would then say, "*describe individual at location A1*"'.

Under no condition can you give any information that is directly pertinent to the target. There is never any front-loading. The reason for this is because the entire concept of remote viewing is that an individual is forced, has no choice, but to use their psi ability to answer the requirement. Any info that is given in any way, or form, modifies that response in a way that removes / reduces the probability of accuracy.

Evidential Details

Beginnings

This details the basis for the original black ops program funding. For readers interested in the data that justified Congressional spending, this initial stage overview of U.S. Military History is recommended.

CIA-Initiated Remote Viewing At Stanford Research Institute

by H. E. Puthoff, Ph.D.[1]
Institute for Advanced Studies at Austin
4030 Braker Lane W., #300
Austin, Texas 78759-5329

Abstract - In July 1995 the CIA declassified, and approved for release, documents revealing its sponsorship in the 1970s of a program at Stanford Research Institute in Menlo Park, CA, to determine whether such phenomena as remote viewing "might have any utility for intelligence collection" [1]. Thus began disclosure to the public of a two-decade-plus involvement of the intelligence community in the investigation of so-called para-psychological or psi phenomena. Presented here by the program's Founder and first Director (1972 - 1985) is the early history of the program, including discussion of some of the first, now declassified, results that drove early interest.

---

[1] Harold Puthoff received his BS and MS Degrees in Electrical Engineering at the University of Florida and a PhD from Stanford University in 1967. He went on to work at the National Security Agency at Fort Meade, Maryland as an Army engineer studying, lasers, high-speed computers, and fiber optics. He is the inventor of the tunable infra-red laser. He spent three years as a naval officer and worked eight years in the Microwave Laboratory at Stanford. Puthoff has over 31 technical papers published on such topics as electron-beam devices, lasers and quantum zero-point-energy effects. He reportedly has patents issued in the areas of energy fields, laser, and communications. [author]

Beginnings

Introduction

On April 17, 1995, President Clinton issued Executive Order Nr. 1995-4-17, entitled Classified National Security Information. Although in one sense the order simply reaffirmed much of what has been long-standing policy, in another sense there was a clear shift toward more openness. In the opening paragraph, for example, we read: "In recent years, however, dramatic changes have altered, although not eliminated, the national security threats that we confront. These changes provide a greater opportunity to emphasize our commitment to open Government." In the Classification Standards section of the Order this commitment is operationalized by phrases such as "If there is significant doubt about the need to classify information, it shall not be classified." Later in the document, in reference to information that requires continued protection, there even appears the remarkable phrase "In some exceptional cases, however, the need to protect such information may be outweighed by the public interest in disclosure of the information, and in these cases the information should be declassified."

A major fallout of this reframing of attitude toward classification is that there is enormous pressure on those charged with maintaining security to work hard at being responsive to reasonable requests for disclosure. One of the results is that FOIA (Freedom of Information Act) requests that have languished for months to years are suddenly being acted upon.[1]

One outcome of this change in policy is the government's recent admission of its two-decade-plus involvement in funding highly-classified, special access programs in remote viewing (RV) and related psi phenomena, first at Stanford Research Institute (SRI) and then at Science Applications International Corporation (SAIC), both in Menlo Park, CA, supplemented by various in-house government programs. Although almost all of the documentation remains yet classified, in July 1995 270 pages of SRI reports were declassified and released by the CIA, the program's first sponsor [2]. Thus, although through the years columns by Jack Anderson and others had claimed leaks of "psychic spy" programs with such exotic names as Grill Flame, Center Lane, Sunstreak and Star

Evidential Details

Gate, CIA's release of the SRI reports constitutes the first documented public admission of significant intelligence community involvement in the psi area.

As a consequence of the above, although I had founded the program in early 1972, and had acted as its Director until I left in 1985 to head up the Institute for Advanced Studies at Austin (at which point my colleague Ed May assumed responsibility as Director), it was not until 1995 that I found myself for the first time able to utter in a single sentence the connected acronyms CIA/SRI/RV. In this report I discuss the genesis of the program, report on some of the early, now declassified, results that drove early interest, and outline the general direction the program took as it expanded into a multi-year, multi-site, multi-million-dollar effort to determine whether such phenomena as remote viewing "might have any utility for intelligence collection" [1].

Beginnings

In early 1972, I was involved in laser research at Stanford Research Institute (now called SRI International) in Menlo Park, CA. At that time I was also circulating a proposal to obtain a small grant for some research in quantum biology. In that proposal I had raised the issue whether physical theory as we knew it was capable of describing life processes, and had suggested some measurements involving plants and lower organisms [3]. This proposal was widely circulated, and a copy was sent to Cleve Backster in New York City who was involved in measuring the electrical activity of plants with standard polygraph equipment. New York artist Ingo Swann chanced to see my proposal during a visit to Backster's lab, and wrote me suggesting that if I were interested in investigating the boundary between the physics of the animate and inanimate, I should consider experiments of the parapsychological type. Swann then went on to describe some apparently successful experiments in psychokinesis in which he had participated at Prof. Gertrude Schmeidler's laboratory at the City College of New York. As a result of this correspondence I invited him to visit SRI for a week in June 1972 to demonstrate such effects, frankly, as much out of personal scientific curiosity as

Beginnings

anything else.

Prior to Swann's visit I arranged for access to a well-shielded magneto-meter used in a quark-detection experiment in the Physics Department at Stanford University. During our visit to this laboratory, sprung as a surprise to Swann, he appeared to perturb the operation of the magnetometer, located in a vault below the floor of the building and shielded by mu-metal shielding, an aluminum container, copper shielding and a superconducting shield. As if to add insult to injury, he then went on to "remote view" the interior of the apparatus, rendering by drawing a reasonable facsimile of its rather complex (and heretofore unpublished) construction. It was this latter feat that impressed me perhaps even more than the former, as it also eventually did representatives of the intelligence community. I wrote up these observations and circulated it among my scientific colleagues in draft form of what was eventually published as part of a conference proceeding [4].

In a few short weeks a pair of visitors showed up at SRI with the above report in hand. Their credentials showed them to be from the CIA. They knew of my previous background as a Naval Intelligence Officer and then civilian employee at the National Security Agency (NSA) several years earlier, and felt they could discuss their concerns with me openly. There was, they told me, increasing concern in the intelligence community about the level of effort in Soviet parapsychology being funded by the Soviet security services [5]; by Western scientific standards the field was considered nonsense by most working scientists. As a result they had been on the lookout for a research laboratory outside of academia that could handle a quiet, low-profile classified investigation, and SRI appeared to fit the bill. They asked if I could arrange an opportunity for them to carry out some simple experiments with Swann, and, if the tests proved satisfactory, would I consider a pilot program along these lines? I agreed to consider this, and arranged for the requested tests. [2]

The tests were simple, the visitors simply hiding objects in a box and asking Swann to attempt to describe the contents. The results generated in these experiments are perhaps captured most eloquently by the following example. In one test Swann said "I see something small, brown and irregular, sort of like a leaf or something that resembles it, except that it seems very much alive, like it's even moving!" The target chosen by one of the visitors turned out to be a small live moth, which indeed did look like a leaf. Although not all responses were quite so precise, nonetheless the integrated results were sufficiently impressive that in short order an eight-month, $49,909 Biofield Measurements Program was negotiated as a pilot study, a laser colleague Russell Targ who had had a long-time interest and involvement in parapsychology joined the program, and the experimental effort was begun in earnest.

Early Remote Viewing Results

During the eight-month pilot study of remote viewing the effort gradually evolved from the remote viewing of symbols and objects in envelopes and boxes, to the remote viewing of local target sites in the San Francisco Bay area, demarked by outbound experimenters sent to the site under strict protocols devised to prevent artifactual results. Later judging of the results were similarly handled by double-blind protocols designed to foil artifactual matching. Since these results have been presented in detail elsewhere, both in the scientific literature [6-8] and in popular book format [9], I direct the interested reader to these sources. To summarize, over the years the back-and-forth criticism of protocols, refinement of methods, and successful replication of this type of remote viewing in independent laboratories [10-14], has yielded considerable scientific evidence for the reality of the phenomenon. Adding to the strength of these results was the discovery that a growing number of individuals could be found to demonstrate high-quality remote viewing, often to their own surprise, such as the talented Hella Hammid. As a separate issue, however, most convincing to our early program monitors were the results now to be described, generated under their own control.

First, during the collection of data for a formal remote viewing series targeting indoor laboratory apparatus and outdoor locations (a series eventually published in toto in the Proc. IEEE [7]), the CIA contract monitors, ever watchful for possible

Beginnings

chicanery, participated as remote viewers themselves in order to critique the protocols. In this role three separate viewers, designated visitors V1 - V3 in the IEEE paper, contributed seven of the 55 viewings, several of striking quality. Reference to the IEEE paper for a comparison of descriptions/ drawings to pictures of the associated targets, generated by the contract monitors in their own viewings, leaves little doubt as to why the contract monitors came to the conclusion that there was something to remote viewing (see, for example, Figure 1 herein).

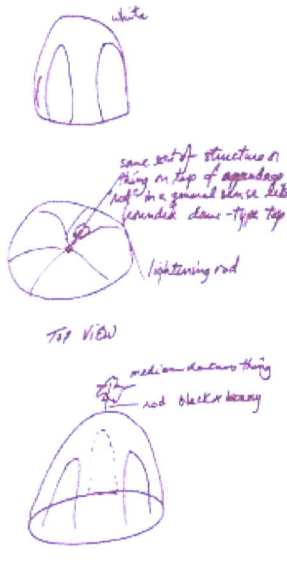

Figure 1 – Sketch of target by VI

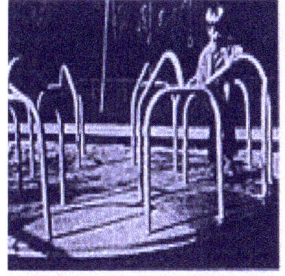

Figure 2 - Target (merry-go-round)

As summarized in the Executive Summary of the now-released Final Report [2] of the second year of the program, "The development of this capability at SRI has evolved to the point where visiting CIA personnel with no previous exposure to such concepts have performed well under controlled laboratory conditions (that is, generated target descriptions of sufficiently high quality to permit blind matching of descriptions to targets by independent judges)." What happened next, however, made even these results pale in comparison.

Coordinate Remote Viewing

To determine whether it was necessary to have a "beacon" individual at the target site, Swann suggested carrying out an experiment to remote view the planet Jupiter before the upcoming NASA Pioneer 10 fly by. In that case, much to his chagrin (and ours) he found a ring around Jupiter, and wondered if perhaps he had remote viewed Saturn by mistake. Our colleagues in astronomy were quite unimpressed as well, until the flyby revealed that an unanticipated ring did in fact exist. [3] Expanding the protocols yet further, Swann proposed a series of experiments in which the target was designated not by sending a "beacon" person to the target site, but rather by the use of geographical coordinates, latitude and longitude in degrees, minutes and seconds. Needless to say, this proposal seemed even more outrageous than "ordinary" remote viewing. The difficulties in taking this proposal seriously, designing protocols to eliminate the possibility of a combination of globe memorization and eidetic or photographic memory, and so forth, are discussed in considerable detail in Reference [9]. Suffice it to say that investigation of this approach, which we designated Scanate (scanning by coordinate), eventually provided us with sufficient evidence to bring it up to the contract monitors and suggest a test under their control. A description of that test and its results, carried out in mid-1973 during the initial pilot study, are best presented by quoting directly from the Executive Summary of the Final Report of the second year's follow-up program [2]. The remote viewers were Ingo Swann and Pat Price, and the entire transcripts are available in the released documents [2].

In order to subject the remote viewing phenomena to a rigorous long distance test under external control, a request for geographical coordinates of a site unknown to subject and experimenters was forwarded to the OSI group responsible for threat analysis in this area. In response, SRI personnel received a set of geographical coordinates (latitude and longitude in degrees,

Beginnings

minutes, and seconds) of a facility, hereafter referred to as the West Virginia Site. The experimenters then carried out a remote viewing experiment on a double-blind basis, that is, blind to experimenters as well as subject. The experiment had as its goal the determination of the utility of remote viewing under conditions approximating an operational scenario. Two subjects targeted on the site, a sensitive installation. One subject drew a detailed map of the building and grounds layout, the other provided information about the interior including code words, data subsequently verified by sponsor sources (report available from COTR).[4]

Since details concerning the site's mission in general, [5] and evaluation of the remote viewing test in particular, remain highly classified to this day, all that can be said is that interest in the client community was heightened considerably following this exercise.

Because Price found the above exercise so interesting, as a personal challenge he went on to scan the other side of the globe for a Communist Bloc equivalent and found one located in the Urals, the detailed description of which is also included in Ref. [2]. As with the West Virginia Site, the report for the Urals Site was also verified by personnel in the sponsor organization as being substantially correct.

What makes the West Virginia/Urals Sites viewings so remarkable is that these are not best-ever examples culled out of a longer list; these are literally the first two site-viewings carried out in a simulated operational-type scenario. In fact, for Price these were the very first two remote viewings in our program altogether, and he was invited to participate in yet further experimentation.

Operational Remote Viewing (Semipalatinsk, USSR)

Midway through the second year of the program (July 1974) our CIA sponsor decided to challenge us to provide data on a Soviet site of ongoing operational significance. Pat Price was the remote viewer. A description of the remote viewing, taken from our declassified final report [2], reads as given below. I cite this level of detail to indicate the thought that goes into such an "experiment" to minimize cueing while at the same time being responsive to the

Evidential Details

requirements of an operational situation. Again, this is not a "best-ever" example from a series of such viewings, but rather the very first operational Soviet target concerning which we were officially tasked. "To determine the utility of remote viewing under operational conditions, a long-distance remote viewing experiment was carried out on a sponsor designated target of current interest, an unidentified research center at Semipalatinsk, USSR.

This experiment, carried out in three phases, was under direct control of the COTR. To begin the experiment, the COTR furnished map coordinates in degrees, minutes and seconds. The only additional information provided was the designation of the target as an R&D test facility. The experimenters then closeted themselves with Subject S1, gave him the map coordinates and indicated the designation of the target as an R&D test facility. A remote-viewing experiment was then carried out. This activity constituted Phase I of the experiment.

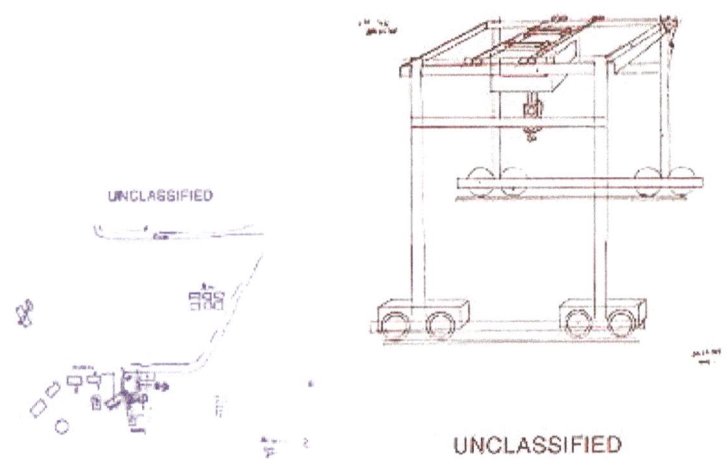

Figure 3 - Subject effort at building layout

Figure 4 - Subject effort construction crane

Figure 3 shows the subject's graphic effort for building layout; Figure 4 shows the subject's particular attention to a multistory gantry crane he observed at the site. Both results were obtained by the experimenters on a double-blind basis before exposure to any additional COTR-held information, thus

Beginnings

eliminating the possibility of cueing. These results were turned over to the client representatives for evaluation. For comparison, an artist's rendering of the site as known to the COTR (but not to the experimenters until later) is shown in Figure 5.

Were the results not promising, the experiment would have stopped at this point. Description of the multistory crane, however, a relatively unusual target item, was taken as indicative of possible target acquisition. Therefore, Phase II was begun, defined by the subject being made "witting" (of the client) by client representatives who introduced themselves to the subject at that point; Phase II also included a second round of experimentation on the Semipalatinsk site with direct participation of client representatives in which further data were obtained and evaluated. As preparation for this phase, client representatives purposely kept themselves blind to all but general knowledge of the target site to minimize the possibility of cueing. The Phase II effort was focused on the generation of physical data that could be independently verified by other client sources, thus providing a calibration of the process.

The end of Phase II gradually evolved into the first part of Phase III, the generation of unverifiable data concerning the Semipalatinsk site not available to the client, but of operational interest nonetheless. Several hours of tape transcript and a notebook of drawings were generated over a two-week period.

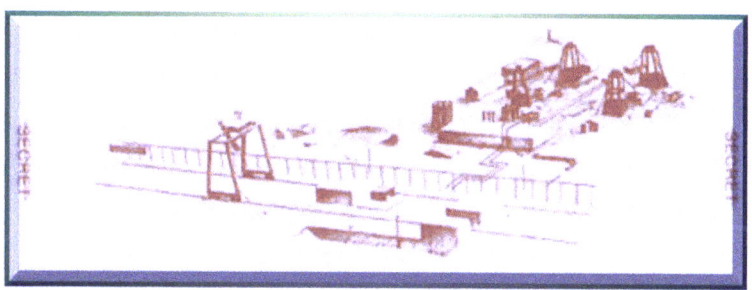

Figure 5 - Actual COTR rendering of Semipalatinsk, USSR target site.

The data describing the Semipalatinsk site were evaluated by the sponsor, and are contained in a separate report. In general, several details concerning the salient technology of the

Semipalatinsk site appeared to dovetail with data from other sources, and a number of specific large structural elements were correctly described. The results contained noise along with the signal, but were nonetheless clearly differentiated from the chance results that were generated by control subjects in comparison experiments carried out by the COTR."

For discussion of the ambiance and personal factors involved in carrying out this experiment, along with further detail generated as Price (see Figure 6) "roamed" the facility, including detailed comparison of Price's RV-generated information with later determined "ground-truth reality," see the accompanying article by Russell Targ in the Journal of Scientific Exploration <http:// www.jse.com/>, Vol. 10, No. 1.

Additional experiments having implications for intelligence concerns were carried out, such as the remote viewing of cipher machine type apparatus, and the RV-sorting of sealed envelopes to differentiate those that contained letters with secret writing from those that did not. To discuss these here in detail would take us too far afield, but the interested reader can follow up by referring to the now-declassified project documents [2].

Follow-on Programs

The above discussion brings us up to the end of 1975. As a result of the material being generated by both SRI and CIA remote viewers, interest in the program in government circles, especially within the intelligence community, intensified considerably and led to an ever increasing briefing schedule. This in turn led to an ever-increasing number of clients, contracts and tasking, and therefore expansion of the program to a multi-client base, and eventually to an integrated joint-services program under single-agency (DIA)[6] leadership. To meet the demand for the increased level of effort we first increased our professional staff by inviting Ed May to join the program in 1976, then screened and added to the program a cadre of remote viewers as consultants, and let subcontracts to increase our scope of activity.

As the program expanded, in only a very few cases could the client's identities and program tasking be revealed. Examples

Beginnings

include a NASA-funded study negotiated early in the program by Russ Targ to determine whether the internal state of an electronic random-number-generator could be detected by RV processes [16], and a study funded by the Naval Electronics Systems Command to determine whether attempted remote viewing of distant light flashes would induce correlated changes in the viewer's brainwave (EEG) production [17]. For essentially all other projects, during my 14-yr. tenure at SRI, however, the identity of the clients and most of the tasking were classified and remain so today. (The exception was the occasional privately funded study.) We are told, however, that further declassification and release of much of this material is almost certain to occur.

What can be said, then, about further development of the program in the two decades following 1975?[7] In broad terms it can be said that much of the SRI effort was directed not so much toward developing an operational U.S. capability, but rather toward assessing the threat potential of its use against the U.S. by others.

The words 'threat assessment' were often used to describe the program's purpose during its development, especially during the early years. As a result much of the remote-viewing activity was carried out under conditions where ground-truth reality was a priori known or could be determined, such as the description of U.S. facilities and technological developments, the timing of rocket test firings and underground nuclear tests, and the location of individuals and mobile units. And, of course, we were responsive to requests to provide assistance during such events as the loss of an airplane or the taking of hostages, relying on the talents of an increasing cadre of remote-viewer/consultants, some well-known in the field such as Keith Harary, and many who have not surfaced publicly until recently, such as Joe McMoneagle.

One might ask whether in this program RV-generated information was ever of sufficient significance as to influence decisions at a policy level. This is of course impossible to determine unless policymakers were to come forward with a statement in the affirmative. One example of a possible candidate is a study we performed at SRI during the Carter administration debates concerning proposed deployment of the mobile MX

Evidential Details

missile system. In that scenario missiles were to be randomly shuffled from silo to silo in a silo field, in a form of high-tech shell game. In a computer simulation of a twenty-silo field with randomly-assigned (hidden) missile locations, we were able, using RV-generated data, to show rather forcefully that the application of a sophisticated statistical averaging technique (sequential sampling) could in principle permit an adversary to defeat the system. I briefed these results to the appropriate offices at their request, and a written report with the technical details was widely circulated among groups responsible for threat analysis [18], and with some impact. What role, if any,

Figure 6 - Left to right: Christopher Green,[2] Pat Price,[3] and Hal Puthoff. Picture taken following a successful experiment involving glider-ground RV.

---

[2] Dr. Christopher Green MD. Neurophysiology, received the CIA's National Intelligence Medal as a Scientific Advisory Board Member to the CIA's Directorate of Intelligence.

[3] One of the finest remote viewers ever, Pat Price, a former police commissioner and councilman in Burbank, CA, came to the Government's attention when he viewed officers, interiors, and files at the virtually unknown, nuclear hardened Naval Satellite Intelligence site in West Virginia. When the Pentagon was shown the data, Price was interrogated by the U.S. Defense Investigative Service who demanded to know who had breached security and how they did it. He is reputed to be the only viewer that could read numbers and letters on a target. Later he viewed inside the Soviet installation at Mount Narodnaya in the Ural Mountains. He went on to work for the CIA and is reputed to have died of a heart attack in July of 1975, in Las Vegas. Even though he was supposedly dead on arrival at the hospital, no autopsy was performed. Suspicions have always existed about the truth of his death. [author]

Beginnings

our small contribution played in the mix of factors behind the enormously complex decision to cancel the program will probably never be known, and must of course a priori be considered in all likelihood negligible. Nonetheless, this is a prototypical example of the kind of tasking that by its nature potentially had policy implications.

Even though the details of the broad range of experiments, some brilliant successes, many total failures, have not yet been released, we have nonetheless been able to publish summaries of what was learned in these studies about the overall characteristics of remote viewing, as in Table 5 of Reference [8]. Furthermore, over the years we were able to address certain questions of scientific interest in a rigorous way and to publish the results in the open literature. Examples include the apparent lack of attenuation of remote viewing due to seawater shielding (submersible experiments) [8], the amplification of RV performance by use of error-correcting coding techniques [19, 20], and the utility of a technique we call associational remote viewing (ARV) to generate useful predictive information [21].8

As a sociological aside, we note that the overall efficacy of remote viewing in a program like this was not just a scientific issue. For example, when the Semipalatinsk data described earlier was forwarded for analysis, one group declined to get involved because the whole concept was unscientific nonsense, while a second group declined because, even though it might be real, it was possibly demonic; a third group had to be found. And, as in the case of public debate about such phenomena, the program's image was on occasion as likely to be damaged by an over enthusiastic supporter, as by a detractor. Personalities, politics and personal biases were always factors to be dealt with.

Official Statements/Perspectives

With regard to admission by the government of its use of remote viewers under operational conditions, officials have on occasion been relatively forthcoming. President Carter, in a speech to college students in Atlanta in September 1995, is quoted by Reuters as saying that during his administration a plane went down

Evidential Details

in Zaire, and a meticulous sweep of the African terrain by American spy satellites failed to locate any sign of the wreckage. It was then "without my knowledge" that the head of the CIA (Adm. Stansfield Turner) turned to a woman reputed to have psychic powers. As told by Carter, "she gave some latitude and longitude figures. We focused our satellite cameras on that point and the plane was there." Independently, Turner himself also has admitted the Agency's use of a remote viewer (in this case, Pat Price).[9] And recently, in a segment taped for the British television series Equinox [22], Maj. Gen. Ed Thompson, Assistant Chief of Staff for Intelligence, U.S. Army (1977-1981), volunteered "I had one or more briefings by SRI and was impressed.... The decision I made was to set up a small, in-house, low-cost effort in remote viewing...."

Finally, a recent unclassified report [23] prepared for the CIA by the American Institutes for Research (AIR), concerning a remote viewing effort carried out under a DIA program called Star Gate (discussed in detail elsewhere in this volume), cites the roles of the CIA and DIA in the history of the program, including acknowledgment that a cadre of full-time government employees used remote viewing techniques to respond to tasking from operational military organizations. [10]

As information concerning the various programs spawned by intelligence-community interest is released, and the dialog concerning their scientific and social significance is joined, the results are certain to be hotly debated. Bearing witness to this fact are the companion articles in this volume by Ed May, Director of the SRI and SAIC programs since 1985, and by Jessica Utts and Ray Hyman, consultants on the AIR evaluation cited above. These articles address in part the AIR study. That study, limited in scope to a small fragment of the overall program effort, resulted in a conclusion that although laboratory research showed statistically significant results, use of remote viewing in intelligence gathering was not warranted.

Regardless of one's a priori position, however, an unimpassioned observer cannot help but attest to the following fact. Despite the ambiguities inherent in the type of exploration

Beginnings

covered in these programs, the integrated results appear to provide unequivocal evidence of a human capacity to access events remote in space and time, however falteringly, by some cognitive process not yet understood. My years of involvement as a research manager in these programs have left me with the conviction that this fact must be taken into account in any attempt to develop an unbiased picture of the structure of reality.

Footnotes

1 - One example being the release of documents that are the subject of this report - see the memoir by Russell Targ.

2 - Since the reputation of the intelligence services is mixed among members of the general populace, I have on occasion been challenged as to why I would agree to cooperate with the CIA or other elements of the intelligence community in this work. My answer is simply that as a result of my own previous exposure to this community I became persuaded that war can almost always be traced to a failure in intelligence, and that therefore the strongest weapon for peace is good intelligence.

3 - This result was published by us in advance of the ring's discovery [9].

4 - Editor's footnote added here: COTR - Contracting Officer's Technical Representative.

5 - An NSA listening post at the Navy's Sugar Grove facility, according to intelligence-community chronicler Bamford [15]

6 - DIA - Defense Intelligence Agency. The CIA dropped out as a major player in the mid-seventies due to pressure on the Agency (unrelated to the RV Program) from the Church-Pike Congressional Committee.

7 - See also the contribution by Ed May elsewhere in this volume concerning his experiences from 1985 on during his tenure as Director.

8 - For example, one application of this technique yielded not only a published, statistically significant result, but also a return of $26,000 in 30 days in the silver futures market [21].

9 - The direct quote is given in Targ's contribution elsewhere in this volume.

Evidential Details

10 - "From 1986 to the first quarter of FY 1995, the DoD paranormal psychology program received more than 200 tasks from operational military organizations requesting that the program staff apply a paranormal psychological technique know (sic) as "remote viewing" (RV) to attain information unavailable from other sources." [23]

References

[1] "*CIA Statement on 'Remote Viewing*," CIA Public Affairs Office, 6 September 1995.

[2] Harold E. Puthoff and Russell Targ, "*Perceptual Augmentation Techniques,*" SRI Progress Report No. 3 (31 Oct. 1974) and Final Report (1 Dec. 1975) to the CIA, covering the period January 1974 through February 1975, the second year of the program. This effort was funded at the level of $149,555.

[3] H. E. Puthoff, "*Toward a Quantum Theory of Life Process*," unpubl proposal, Stanford Research Institute (1972).

[4] H. E. Puthoff and R. Targ, "*Physics, Entropy and Psycho-kinesis,*" in Proc. Conf. Quantum Physics and Parapsychology (Geneva, Switzerland); (New York: Parapsychology Foundation, 1975).

[5] Documented in "*Paraphysics R&D - Warsaw Pact* (U)," DST-1810S-202-78, Defense Intelligence Agency (30 March 1978).

[6] R. Targ and H. E. Puthoff, "*Information Transfer under Conditions of Sensory Shielding,*" Nature 252, 602 (1974).

[7] H. E. Puthoff and R. Targ, "*A Perceptual Channel for Information Transfer over Kilometer Distances: Historical Perspective and Recent Research,*" Proc. IEEE 64, 329 (1976).

[8] H. E. Puthoff, R. Targ and E. C. May, "*Experimental Psi Research: Implications for Physics,*" in The Role of Consciousness in the Physical World", edited by R. G. Jahn (AAAS Selected Symposium 57, Westview Press, Boulder, 1981).

[9] R. Targ and H. E. Puthoff, *Mind Reach* (Delacorte Press, New York, 1977).

[10] J. P. Bisaha and B. J. Dunne, "*Multiple Subject and Long-Distance Precognitive Remote Viewing of Geographical Locations,*" in Mind at Large, edited by C. T. Tart, H. E. Puthoff and R. Targ (Praeger, New York, 1979), p. 107.

[11] B. J. Dunne and J. P. Bisaha, "*Precognitive Remote Viewing in the Chicago Area: a Replication of the Stanford Experiment,*" J. Parapsychology 43, 17 (1979).

Beginnings

[12] R. G. Jahn, "*The Persistent Paradox of Psychic Phenomena: An Engineering Perspective*," Proc. IEEE 70, 136 (1982).
[13] R. G. Jahn and B. J. Dunne, "*On the Quantum Mechanics of Consciousness with Application to Anomalous Phenomena*," Found. Phys. 16, 721 (1986).
[14] R. G. Jahn and B. J. Dunne, *Margins of Reality* (Harcourt, Brace and Jovanovich, New York, 1987).
[15] J. Bamford, *The Puzzle Palace* (Penguin Books, New York, 1983) pp. 218-222.
[16] R. Targ, P. Cole and H. E. Puthoff, "*Techniques to Enhance Man/Machine Communication,*" Stanford Research Institute Final Report on NASA Project NAS7-100 (August 1974).
[17] R. Targ, E. C. May, H. E. Puthoff, D. Galin and R. Ornstein, "*Sensing of Remote EM Sources* (Physiological Correlates)," SRI Intern'l Final Report on Naval Electronics Systems Command Project N00039-76-C-0077, covering the period November 1975 - to October 1976 (April 1978).
[18] H. E. Puthoff, "*Feasibility Study on the Vulnerability of the MPS System to RV Detection Techniques*," SRI Internal Report, 15 April 1979; revised 2 May 1979.
[19] H. E. Puthoff, "*Calculator-Assisted Psi Amplification*," Research in Parapsychology 1984, edited by Rhea White and J. Solfvin (Scarecrow Press, Metuchen, NJ, 1985), p. 48.
[20] H. E. Puthoff, "*Calculator-Assisted Psi Amplification II: Use of the Sequential-Sampling Technique as a Variable-Length Majority-Vote Code,*" Research in Parapsychology 1985, edited by D. Weiner and D. Radin (Scarecrow Press, Metuchen, NJ, 1986), p. 73.
[21] H. E. Puthoff, "*ARV (Associational Remote Viewing) Applications,*" Research in Parapsychology 1984, edited by Rhea White and J. Solfvin (Scarecrow Press, Metuchen, NJ, 1985), p. 121.
[22] "*The Real X-Files*", Independent Channel 4, England (shown 27 August 1995); to be shown in the U.S. on the Discovery Channel.
[23] M. D. Mumford, A. M. Rose and D. Goslin, "*An Evaluation of Remote Viewing: Research and Applications*", American Institutes for Research (September 29, 1995).

Copyright 1996 by Dr. H.E. Puthoff.

Permission to redistribute granted, but only in complete and unaltered form.

[The footnotes are designed to facilitate a greater understanding of Remote Viewing pioneers, but are not original. None of Dr. Puthoff's text was altered.]

Evidential Details

Targeted Reading

Because of its capabilities Remote Viewing disinformation exists which discourages further interest. This list was assembled to help people locate books directly from members of the program involved in this most fascinating component of United States Military History.

Books by Members of the U.S. Military Program

McMoneagle, Joseph W.
- *Mind Trek*; Hampton Roads, 1993
- *The Ultimate Time Machine*; Hampton Roads, 1998
- *Remote Viewing Secrets*; Hampton Roads, 2000
- *The Stargate Chronicles*; Hampton Roads, 2002
- *Memoirs of a Psychic Spy: The Remarkable Life of U. S. Government Remote Viewer 001*; Hampton Roads, 2006

Buchanan, Leonard
- *The Seventh Sense – The Secrets of Remote Viewing as Told by a "Psychic Spy" for the U.S. Military*; Paraview Pocket Books, 2003
- *Remote Viewing Methods - Remote Viewing and Remote Influencing*; DVD, 2004

Smith, Paul H.
- *Reading the Enemy's Mind - Inside Stargate - America's Psychic Espionage Program*; Tor non-fiction, 2005

Morehouse, David A.
- *Psychic Warrior – Inside the CIA's Stargate Program: The True Story of a Soldiers Espionage and Awakening*; St Martin's Press, 1996
- *Nonlethal Weapons: War Without Death*; Praeger Publishers, 1996
- *Remote Viewing: The Complete User's Manual for Coordinate Remote Viewing*; Sounds True Publishers, 2011

Puthoff, Harold E. with Russell Targ
- *Mind Reach - Scientists Look at Psychic Abilities*; Delacorte, 1977 & New World Library, 2004

Swann, Ingo
- *To Kiss the Earth Goodbye*; Hawthorne, New York, 1975
- *Star Fire*, Dell non-fiction, 1978
- *Natural ESP: The ESP Core and its Raw Characteristics* with Harold E. Puthoff; Bantam Books, 1987

Targeted Reading

- *Everybody's Guide to Natural ESP: Unlocking The Extrasensory Power of Your Mind*; Jeremy P. Tharcher Imprint, 1991
- *Your Nostradamus Factor*; Fireside Press, 1993
- *Remote Viewing & ESP From The Inside Out*; DVD

Targ, Russell
- *Mind Race: Understanding and Using Psychic Abilities*, with Keith Harary; Ballantine Books, 1984
- *Miracles of Mind: Exploring Nonlocal Consciousness and Spiritual Healing*; New World Library, 1999
- *Limitless Mind: A Guide to Remote Viewing and Transformation of Consciousness*; New World Library, 2004
- *Do you See What I See?*; *ESP and the C.I.A. and the Meaning of Life*; Hampton Roads, 2010
- *The Reality of ESP: A Physicists Proof of Psychic Abilities*; Quest Books, 2012

Atwater, F. Holmes
- *Captain of My Ship, Master of My Soul: Living with Guidance*; Hampton Roads Publishing, 2001

Other Sources

Monroe, Robert
- *Journeys Out of the Body*; Three Rivers Press, 1992
- *Ultimate Journey*; Three Rivers Press, 1996

Radin, Dean I.
- *The Conscious Universe: The Scientific Truth of Psychic Phenomena*; Harper Edge, 1997
- *Entangled Minds: Extrasensory Experiences in a Quantum Reality*, Paraview Pocket Books, 2006

- Moreno, Jonathon D. - *Mind Wars: Brain Science and the Military in the 21st Century*; Bellevue Literary Press, 2012

- Schnabel, Jim – *Remote Viewers: The Secret History of America's Psychic Spies*; Dell–non-fiction, 1997

- McRae, Ronald – *Mind Wars: The true story of Government Research into the Military Potential of Psychic Weapons*; St Martin's Press, 1984

Evidential Details

Additional Taskings

Lae City Airport, New Guinea - July, 1937 – Get into the cockpit for the last flight of the vanished pilot **Amelia Earhart**. Learn of the plane's unknown final flight trajectory, cockpit circumstances and final thoughts in her last minute of life. Entered into four libraries within six months, including Purdue University's Earhart Special Collection Library, the book includes a "how to find the debris field" location map with yardages and points of reference including a perfect flight scenario that has never been put forward. With the continuous failures of others, insiders have subsequently blogged that our scenario is the one worth pursuing.

Ötzal Alps Mountains - Italian-Austrian border ~ 3,300 BC – Follow the trail of Europe's archeological "show of the century". Learn the whereabouts of **Ötzi the Iceman**'s unknown home camp and why and how he died alone in the mountains which some falsely regard as a Neolithic crime scene. This book includes remote viewing maps, pre-death tool drawings, including an undiscovered artifact, his cabin, and the world's only real time portrait considered significant enough that the Museum in Bolzano, Italy obtained its copywrite release for Ötzi's 20th Anniversary exhibit. Interwoven with scientific quotation, this account also includes specifics of his tribal life in what we first identified as the Langtaufers Valley. The book provides Ötzi's previously unknown eight day course through the mountains using modern Alpine trail numbers. Learn the real reason for his violent death as he was alone at 10,500 feet in the Otzal Alps.

Onboard RMS Titanic - North Atlantic - April, 1912 – A griping story, with select court room testimony, substantiating our strange crow's nest events as *Titanic* bore down on the ice. Then, move to a resolution of **Captain E. J. Smith**'s final actions in his previously unknown non-drowning death. The book includes artifact drawings whose existence was only confirmed through ocean floor salvage after the remote viewing sessions. Once the last lifeboat was away, those left behind knew death was imminent. Read

Additional Taskings

History's only narrative of the last wild 20 minutes as the ship prepared to take over 1500 terrified travelers into the frigid black ocean at 2:18 in the morning. No other book has all this.

Last Stand Hill - Little Big Horn River - Montana - June, 1876 – This is history's only documentation of General **George Armstrong Custer**'s last stand from the viewpoints of the victors and the vanquished. Read about Chief Sitting Bull's as well as Custer's battle thoughts. Learn of the true cause of Custer's death and the amazing reasons his body is likely not in his tomb at West Point. You get new, remote viewing generated, battle maps with a drawings of Custer's last fighting stance, a near death facial close-up and, since he was never photographed, the world's only full page color portrait of Indian War Chief *Crazy Horse*. Completely unique.

Execution Square - Rouen, France - May, 1431 – Go to stake in the market square for the execution of the military heroine lost in the mists of time - **Joan of Arc**. Recounted are her military successes, capture, the political intrigues and some excerpts from her heresy trial. Study an amazingly detailed medieval architectural description of Rouen's town square as Joan saw it. McMoneagle's renowned artwork documents the scene as she was chained to the burning scaffold (not a stake). The Evidential Details prove she did not die by flame. The book includes the **world's only portrait** of the previously faceless heroine who went on to become a Saint.

Each edition in the Evidential Details Mystery Series is all you need to learn the how, when's and whys of each mystery with a "you are there" first person viewpoint. Sit back and enjoy these first time ever books using the number 1 Operation Star Gate viewer.

Evidential Details

Historians wage a constant battle in trying to pin down reality. How they are being treated by the political, social, or religious pressures in the now of their time and place will usually dictate how they will report reality. Since it is the political, social, and religious environment that pays for their work, endorses their findings, and passes judgment on them, who can blame for doing otherwise.

~Joseph W. McMoneagle – *The Ultimate Time Machine*~

www.ingramcontent.com/pod-product-compliance
Lightning Source LLC
Chambersburg PA
CBHW040322300426
44112CB00020B/2837